SpringerBriefs in Mathematical Physics

Volume 15

Karen Yeats

A Combinatorial Perspective
on Quantum Field Theory

 Springer

Karen Yeats
Department of Mathematics
Simon Fraser University
Burnaby, BC
Canada

and

Department of Combinatorics
and Optimization
University of Waterloo
Waterloo, ON
Canada

ISSN 2197-1757 ISSN 2197-1765 (electronic)
SpringerBriefs in Mathematical Physics
ISBN 978-3-319-47550-9 ISBN 978-3-319-47551-6 (eBook)
DOI 10.1007/978-3-319-47551-6

Library of Congress Control Number: 2016953661

Printed on acid-free paper

This Springer imprint is published by Springer Nature
The registered company is Springer International Publishing AG
The registered company address is: Gewerbestrasse 11, 6330 Cham, Switzerland

Acknowledgements

I would like to thank all of my colleagues, collaborators, and students, but particularly Dirk Kreimer from whom I learned the keys to this whole area and my students in summer 2016, Iain Crump, Benjamin Moore, Mohamed Laradji, Matthew Lynn, Wesley Chorney, and Maksym Neyra-Nesterenko, who helped proofread this brief. I would also like to thank Cameron Morland for his support.

Contents

Part I
Preliminaries

Chapter 1
Introduction

Quantum field theory is not the first place a combinatorialist is likely to look for interesting problems or to look to apply their techniques. On the other hand, from the physics side, too often combinatorics is viewed as a kind of uninteresting messy detail. However, there is actually a lot of beautiful and useful combinatorics in quantum field theory, and the discrete structures illuminate the physical structure. Neither side is necessarily well positioned to penetrate the literature of the other.

This brief explores combinatorial constructions and discrete-flavoured problems from quantum field theory in a way which is intended to be natural and appealing to a mathematician with a combinatorics background as well as being accessible to mathematical physicists and other mathematicians. It is not comprehensive, but rather takes a tour, shaped by the author's biases, through some of the important ways that a combinatorial perspective can be brought to bear on quantum field theory. In order to retain a strong sense of the overall story and not get lost in the details, the main focus is on giving the objects, constructions, and results in a uniform language, and giving an intuition of why these things are important. Proofs are given when insightful, but others are left to the literature.

This brief has three parts. In the first part the preliminary material will be set out. The second part will discuss Dyson-Schwinger equations. The third part will discuss Feynman graph periods. The second and third parts are largely independent of each other and can be read in either order.

The first stop on this tour will be a rough overview of what quantum field theory is all about. Then we will proceed to set up an enumerative framework in Chap. 3 which will be used to understand Dyson-Schwinger equations. Chapter 4 will introduce the combinatorial Hopf algebras that give an algebraic underpinning to renormalization in quantum field theory and underlie much of the graph-level work in later chapters. The preliminary part of the brief ends with Chap. 5 setting up Feynman graphs from a combinatorial and graph theoretic perspective.

Dyson-Schwinger equations are the quantum analogues of equations of motion and so are physically important. Combinatorially they act as a kind of specification and so have a natural enumerative flavour. The Dyson-Schwinger part of the brief begins by setting up Dyson-Schwinger equations in Chap. 6. Then it proceeds to step

© The Author(s) 2017
K. Yeats, *A Combinatorial Perspective on Quantum Field Theory*,
SpringerBriefs in Mathematical Physics 15, DOI 10.1007/978-3-319-47551-6_1

slowly from the purely combinatorial to the more physical. Chapter 7 reviews results
of Foissy giving a classification of when subalgebras coming from combinatorial
Dyson-Schwinger equations are Hopf. Chapter 8 brings in Feynman rules in their
simplest form with the tree factorial. Chapter 9 surveys results on expressing solutions
to a class of more physical Dyson-Schwinger equations in terms of expansions over
chord diagrams. The Dyson-Schwinger part concludes with Chap. 10 describing
recent results on viewing log expansions with combinatorial tools.

The final part turns to individual Feynman graphs and Feynman integrals with
a focus on a particular renormalization scheme independent residue known as the
Feynman period. Chapter 11 gives the combinatorial and analytic definitions that
will be needed. Next we look at graph symmetries that preserve the Feynman period
in Chap. 12. Chapter 13 then looks at a graph invariant known to have these same
symmetries, but for which it is not known how it relates to the period itself. Chapter 14
introduces Brown's denominator reduction algorithm and discusses its relation to the
weight of the period. The weight of the period leads to an arithmetic graph invariant
known as the c_2 invariant. What we know about the c_2 invariant is reviewed in
Chap. 15. The focus and language remain largely combinatorial. For this invariant
the connection with the period is more clear but the some of the symmetries are
conjectural. Finally, the brief concludes in Chap. 16 with a brief review of some of
the more combinatorial aspects of the Feynman integration algorithms which have
been built around these ideas.

Chapter 2
Quantum Field Theory Set Up

Some standard introductions to quantum field theory are [1–3], for a particularly diagrammatic approach see [4]. For the reader who is not familiar with these ideas we will briefly go over the intuition of what quantum field theory is along with some of the key vocabulary. Many readers would be safe skipping this chapter either because they are familiar with this material or because they are more interested in the problems which appear later than in their motivation.

Quantum field theory is a framework in which we can understand arbitrary numbers of interacting particles quantum mechanically. It is the standard way to unify quantum mechanics and special relativity. The particles in question can be subatomic particles in high energy physics in which case quantum field theory, through the standard model, describes all known particles extremely well. The particles can also be quasiparticles in condensed matter physics and so quantum field theory is a useful tool for understanding condensed matter systems and the mathematician or mathematical physicist gets new theories to play with.

In either case, the fundamental thing a quantum field theory describes is how particles interact and scatter, so one imagines an idealized experiment where some known particles are sent in, collide and interact in some way, and then what comes out is detected. Since we don't know what happened in the collision we, in the spirit of quantum mechanics, take a weighted sum over all possibilities. Any particular story of what the particles did traces out a graph in spacetime with the interactions as vertices and the edges as particles propagating. Combining together those possibilities which after forgetting the spacetime embedding give the same graph, we obtain Feynman graphs,[1] see Fig. 2.1. See Chap. 5 for precise definitions.

The weight of the graph in the sum is its Feynman integral. The weighted sum itself is a *perturbative expansion* for the *scattering amplitude* in question. We'll also see this kind of sum, over appropriate graphs, as *Green functions* when we come to Dyson-Schwinger equations.

Feynman integrals are, in general, very difficult to compute and there is a whole part of high energy physics devoted to the technique and practice of computing

---

[1]Feynman graphs drawn with tikz-feynman [5].

© The Author(s) 2017
K. Yeats, *A Combinatorial Perspective on Quantum Field Theory*,
SpringerBriefs in Mathematical Physics 15, DOI 10.1007/978-3-319-47551-6_2

Fig. 2.1 Example Feynman graphs

them, with the practical aim of computing backgrounds for accelerator experiments and making predictions, see for example the proceedings [6]. For the purposes of this brief there are four things which will be important about Feynman integrals. First, one contribution to the Feynman integral is the strength of each interaction which is captured in one or more ***coupling constants***. The coupling constants can be reinterpreted as counting variables. Second, the Feynman integrand expression can be read off the graph with each edge and vertex contributing a factor. The rules to do this are called ***Feynman rules***. Third, in interesting cases these integrals are divergent and so to extract physically meaningful quantities from them they must be ***renormalized***, see Sect. 4.3 for more on renormalization. Finally, the sums of Feynman integrals contributing to a given process are expected to be divergent for all interesting cases.

From a discrete math perspective, taking a Feynman-graphs-first approach to quantum field theory is quite appealing, as we have graphs playing a central role. Furthermore we have series indexed by graphs which are divergent and hence as a first step are reasonably thought of as formal. There are other less apparent reasons why this is a nice perspective for those with discrete tastes: the structure of the renormalization process is captured with a combinatorial Hopf algebra and important integral and differential equations come from decompositions of combinatorial objects, all of which we will investigate over the course of this brief.

There is a downside to a Feynman-graphs-first approach. The series in question are expected to be divergent in the cases that matter and so they can only be asymptotic series for the presumed functions which describe the physical processes in question. That is, a Feynman-graph-first approach is a perturbative approach. By itself a perturbative approach does not have access to any phenomenon which is asymptotically flat at the point around which we are expanding, that is it cannot see the ***instantons*** in the theory or any other nonperturbative phenomenon. Fortunately, we can access these things by the back door: a Feynman-graphs-first approach doesn't mean a Feynman-graphs-only approach. The way to do this is as follows. The recursive structure of the Feynman graphs and the perturbative expansion give us functional equations for the perturbative expansions. Since these underlying structures are not mere combinatorial happenstance but reflect the physics, they also hold non-perturbatively and so the functional equations can be upgraded to non-perturbative equations where they, potentially at least, can see nonperturbative effects. The functional equations of this type we understand best are Dyson-Schwinger equations. That is why Dyson-Schwinger equations are very important in this approach. To date this is a mere sketch and a lot of work remains before these ideas could be used foundationally for quantum field theory.

More traditionally, quantum field theorists escape the limitations of perturbation theory by beginning with non-perturbative definitions and from there deriving Feynman graphs and the perturbative expansion. One popular and important way to do this is via the path integral, see [3] for an introduction. The initial intuition is very much the same—sum over all possibilities—but here we think of the possibilities as arbitrary paths and so the space of possibilities is continuous and infinite dimensional making the "sum" an integral and, because of the infinite dimensionality, not one which is well defined in general. None-the-less it is an approach which captures the physical intuition well and works in practice, so it's important and interesting even without a complete mathematical foundation.

If spacetime is zero dimensional then the path integral is well defined and we get the zero dimensional field theory approach to counting graphs which is used both by physicists and mathematicians, see for example [7, 8].

In higher dimensions the path integral is still a good candidate for viewing combinatorially simply by temporarily forgetting the analytic difficulties and treating it formally. Jackson, Morales and Kempf have been looking at the enumerative combinatorics of quantum field theory from this perspective. So far this collaboration has resulted in [9, 10] with a comprehensive treatment in the works.

In any case, even purely perturbative quantum field theory is extremely useful and full of interesting mathematics, a small part of which we will investigate in what follows.

References

1. Itzykson, C., Zuber, J.B.: Quantum Field Theory. McGraw-Hill (1980). Dover edition 2005
2. Peskin, M.E., Schroeder, D.V.: An introduction to quantum field theory. Westview (1995)
3. Zee, A.: Quantum field theory in a nutshell. Princeton (2003)
4. Cvitanović, P.: Field Theory. Nordita Lecture Notes (1983)
5. Ellis, J.: Tikz-feynman: Feynman diagrams with tikz. arXiv:1601.05437
6. Blümlein, J., Marquard, P., Riemann, T. (eds.): Loops and Legs in Quantum Field Theory, vol. PoS(LL2014). PoS (2014)
7. Cvitanović, P., Lautrup, B., Pearson, R.B.: Number and weights of Feynman diagrams. Physical Review D **18**(6), 1939–1949 (1978)
8. Lando, S.K., Zvonkin, A.K.: Graphs on Surfaces and Their Applications. Springer (2004)
9. Kempf, A., Jackson, D.M., Morales, A.H.: New Dirac delta function based methods with applications to perturbative expansions in quantum field theory. Journal of Physics A: Mathematical and Theoretical **47**(41), 415,204 (2014). arXiv:1404.0747
10. Kempf, A., Jackson, D.M., Morales, A.H.: How to (path-) integrate by differentiating. Journal of Physics: Conference Series **626**, 012,015 (2015). arXiv:1507.04348

Chapter 3
Combinatorial Classes and Rooted Trees

Throughout this brief, we will use K for the base field and assume that the characteristic is 0. In fact the characteristic restriction is not always necessary and furthermore much of the work could take place over any integral domain or even any commutative ring. This is actually quite typical of combinatorial Hopf algebras, as Grinberg and Reiner [1] have commented, and in particular $K = \mathbb{Z}$ is often useful. However, we will stick to the field case so as to avoid algebraic digressions.

3.1 Combinatorial Classes and Augmented Generating Functions

This section gives an overview of combinatorial classes and their generating functions. A good reference for combinatorial classes in a similar language to the one used here is [2].

Definition 1 A *combinatorial class* $\mathscr{C}$ is a set (by abuse of notation also called $\mathscr{C}$) and a size function $\| \cdot \| : \mathscr{C} \to \mathbb{Z}_{\geq 0}$ with the property that the sets $\mathscr{C}_n = \{c \in \mathscr{C} : |c| = n\}$ are all finite.

The most important combinatorial classes for us will be rooted tree classes. Rooted trees can be defined in many equivalent ways.

One way to define a *rooted tree* is as a finite graph which is connected, has no cycles, and has a distinguished vertex called the root. Given a non-root vertex v, there is a unique vertex adjacent to v and closer to the root than v. This vertex is called v's *parent*. Those vertices (if any) with v as their parent are v's *children*. Mixing metaphors, as is standard, vertices with no children are known as *leaves*. Given a vertex v of a rooted tree. The subtree consisting of v and all its children, all their children, and so on is called the *subtree rooted at* v and will be denoted t_v. We will draw rooted trees with the root at the top.

© The Author(s) 2017
K. Yeats, *A Combinatorial Perspective on Quantum Field Theory*,
SpringerBriefs in Mathematical Physics 15, DOI 10.1007/978-3-319-47551-6_3

Another equivalent way to define a rooted tree is as a finite partially ordered set (poset) with the root as the unique largest element and where every element other than the root has exactly one element covering (i.e. immediately above) it. An ***antichain*** is a subset of the elements of a poset with the property that no element of the antichain is larger than any other element of the antichain; they are all incomparable.

Whichever way one thinks about it, rooted trees form a combinatorial class where the size is the number of vertices. It is sometimes useful to allow an object of size 0 in combinatorial classes of rooted trees. This we call the empty tree, denoted $\mathbb{I}$.

We can form other interesting combinatorial classes by either restricting the trees, say by restricting the number of children vertices can have, or by putting on additional structure. The most important example of additional structure is when we give an ordering to the children at each vertex, resulting in what are called ***plane rooted trees***. We can also form combinatorial classes by describing how to build the elements, such as by giving a combinatorial specification, see Sect. 3.2. This is closely connected to Dyson-Schwinger equations.

Other combinatorial classes which will be very important for us are classes of Feynman graphs, see Chap. 5.

It will be convenient for Part II to take a slightly nonstandard approach to generating functions. First note that any combinatorial class $\mathscr{C}$ can be made into an algebra simply by taking the polynomial algebra $K[\mathscr{C}]$ with generators the elements of $\mathscr{C}$. Addition is purely formal—a sum of trees is just a sum of trees, it is not identified with some other tree or other object. If $\mathscr{C}$ is a class of connected objects, then it will usually make sense to identify multiplication in the polynomial algebra with disjoint union, so that a monomial of elements of $\mathscr{C}$ is the disconnected object given by the disjoint union of the elements. The empty object $\mathbb{I}$ will typically be identified with $1 \in K$.

For example, let $\mathscr{T}$ be the class of rooted trees with no order information at the vertices. Then

$$5\,\overset{\bullet\bullet}{\underset{\bullet}{|}} - 7\,\overset{\bullet\;\bullet}{\wedge} \in K[\mathscr{T}]$$

and we can think of

as a forest of size 5 with two trees.

Now we can define generating functions which keep the objects in the sums.

Definition 2 Given a combinatorial class $\mathscr{C}$, the ***augmented generating function*** of $\mathscr{C}$ is the formal power series

$$C(x) = \sum_{c \in \mathscr{C}} c\, x^{|c|} \in (K[\mathscr{C}])[[x]].$$

For example if we again let $\mathcal{T}$ be the class of rooted trees with no order information at the vertices and say we also include the empty tree in this class, then the augmented generating function of the class begins.

$$T(x) = \mathbb{I} + \bullet x + \mathop{\bullet}^{\bullet} x^2 + \left(\bigwedge + \mathop{\bullet}^{\bullet}_{\bullet} \right) x^3 + \left(\mathop{\bullet}^{\bullet}_{\bullet} + \bigwedge + \mathop{\bullet}^{\bullet}_{\bullet} + \bigwedge \right) x^4 + O(x^5)$$

The next thing we need is an **evaluation map** $\phi : K[\mathscr{C}] \to A$ where A is some algebraic structure over K. Rational functions over K are often a useful choice for A as are formal Laurent series, though, simply to illustrate the underlying mathematical structure, polynomials often suffice.

The simplest evaluation map is defined by $\mathrm{or}(c) = 1$ for all $c \in \mathscr{C}$ and extended as an algebra homomorphism to $K[\mathscr{C}]$. Using this evaluation map on the augmented generating function gives the **ordinary generating function**

$$\sum_{c \in \mathscr{C}} x^{|c|} = \mathrm{or}(C(x)).$$

For example, continuing with $T(x)$ as in the previous example, we have $\mathrm{or}(T(x)) = 1 + x + x^2 + 2x^3 + 4x^4 + O(x^5)$.

Since K has characteristic 0 we have factorials in its field of fractions, so we can define the evaluation map $\mathrm{ex}(c) = 1/|c|!$ for $c \in \mathscr{C}$ and extended as an algebra homomorphism to $K[\mathscr{C}]$. This gives the **exponential generating function**

$$\sum_{c \in \mathscr{C}} \frac{x^{|c|}}{|c|!} = \mathrm{ex}(C(x)).$$

For example, again with the same $T(x)$, we have $\mathrm{ex}(T(x)) = 1 + x + \frac{1}{2}x^2 + \frac{1}{3}x^3 + \frac{1}{6}x^4 + O(x^5)$.

Many examples of multivariate generating functions also fit into this framework. Take one of the variables as the primary variable then the evaluation map will take c to the monomial given by the other variables as they count the parameters of c. For example, suppose we want to make a multivariate generating function for a class of trees where x counts the number of vertices and y counts the number of leaves, then we can use the evaluation map $t \mapsto y^{\text{number of leaves of } t}$.

The example which matters in quantum field theory also fits into this framework. Here the evaluation map is the **Feynman rules** (see Sect. 5.6). This evaluation map will take a Feynman graph, viewed as a combinatorial object (see Chap. 5), and return its Feynman integral. Typically, thinking of a regularized Feynman integral, we could view this map as taking values in the algebra of Laurent expansions in the regularization parameter with the coefficients being functions in the masses of the particles and the kinematical parameters. Alternately, we could view the original Feynman integral as a formal integral expression, and so the Feynman rules take values in some space of formal integral expressions. The latter is the approach taken in subsection 2.3.2 of [3] (or [4]), which additionally discusses treating the renormalized integral formally. The result of evaluating an augmented generating function by Feynman rules will be called a ***Green function***.

We often use rooted trees in place of Feynman diagrams. These trees represent the divergence structure of Feynman diagrams. As we'll see in more detail in Sect. 4.3 and Chap. 5, many Feynman integrals are divergent integrals. We will call a Feynman graph ***divergent*** if it has a divergent Feynman integral. A Feynman graph may also contain proper subgraphs which are divergent. A divergent graph with no divergent subgraphs is called ***primitive***. For renormalization it is very important to understand how divergent subgraphs lie within a Feynman graph—this is the divergence structure of the graph. We often represent this with rooted trees called ***insertion trees***. For example

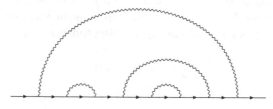

has insertion tree

where the divergent subgraphs are the two copies of

,

the

,

and the whole graph. Which subgraphs are divergent and hence which we take in constructing the tree depends on the physical theory, see Chap. 5 for details. Note however that not all the primitive divergent subgraphs need to be the same, as they were in the case above, and some may be of higher loop order.[1] If we consider the size of a Feynman graph to be its loop order then primitive divergent graphs exist at all loop orders in interesting theories and so to read the size off the tree we should weight the vertices by the loop order of the inserted graphs.

---

[1]The ***loop order***, rephrased in graph theory language, is the dimension of the cycle space of the graph, see Sect. 5.5. In topological language this is the *first Betti number*. *Loop* in Feynman diagram language means ***cycle*** in graph theory language; the graph theorist's loops are called ***tadpoles*** or self-loops.

Note in particular, these trees are not Feynman diagrams themselves; they do not represent tree-level processes, rather there are at least as many loops as vertices, possibly more due to higher loop order primitives.

Also important is that not all Feynman graphs have a tree-like structure to their subdivergences. This phenomenon is known as **overlapping subdivergences**. Fortunately trees still make a good model because we can simply take a sum of trees each representing different ways to resolve the overlaps, see [5]. This works in practice and also has good theoretical justification because of the universality of rooted trees, see Sect. 4.4. An alternate approach is to look at the lattice structure of subdivergences, see Sect. 13 of [6] as well as [7].

The lattice approach is mathematically very pleasing for a few reasons. First of all it honestly captures the structure of subdivergences without any algebraic fudging via universality. Second, it puts renormalization Hopf algebras of Feynman graphs, which will be defined in Chap. 5, into the framework of incidence Hopf algebras [8] which are an important, quite general, and well studied family of combinatorial Hopf algebras. Third the lattice approach can see something about why certain types of graphs are special in quantum field theory, see [7].

None-the-less, rooted trees are an excellent first model for the combinatorics of quantum field theory and we will work with them extensively in the second part of this brief.

We can make toy Feynman rules directly for trees. The simplest cases are just weightings with appropriate algebraic and combinatorial properties. Define the **tree factorial** to be

$$t! = \prod_{v \in t} |t_v|.$$

For example,

$$\text{⋏̣}! = 4 \cdot 2 \cdots 1 \cdot 1 = 8.$$

Then the **tree factorial Feynman rules** are

$$t \mapsto \frac{z^{|t|}}{t!} \tag{3.1}$$

This weighting for trees is important as pure combinatorics since a classical result is that it counts increasing trees. Specifically it is an exercise in Knuth [9] that $|t|!/t!$ counts the number of ways to label the vertices of a plane tree t with $\{1, 2, \ldots, |t|\}$ so that the labels increase from parent to child. On the physics side this example is important because it has the same leading behaviour as realistic Feynman rules, see [10].

We'll see more about tree Feynman rules in Chaps. 4 and 8 and in Chaps. 5 and 11 we'll learn more about physical Feynman rules on Feynman graphs.

3.2   Combinatorial Specifications and Combinatorial Dyson-Schwinger Equations

To give useful specifications of combinatorial classes, we need a collection of combinatorial operations. The operations then translate into operations on the generating functions and so the specifications translate into functional equations satisfied by the generating functions. There are a few competing schools of thought on notation and setup for these operations; this notation is inspired by [2].

The two most basic operations are $+$ and $\times$.

Definition 3 Let $\mathscr{C}$ and $\mathscr{D}$ be two combinatorial classes.

The combinatorial class $\mathscr{C} + \mathscr{D}$ is the disjoint union of $\mathscr{C}$ and $\mathscr{D}$ with the size of an element of $\mathscr{C} + \mathscr{D}$ being its size in $\mathscr{C}$ or $\mathscr{D}$.

The combinatorial class $\mathscr{C} \times \mathscr{D}$ is the Cartesian product of $\mathscr{C}$ and $\mathscr{D}$ with the size of an element of $\mathscr{C} \times \mathscr{D}$ being the sum of the sizes of its $\mathscr{C}$ and $\mathscr{D}$ parts.

We will continue to use $\mathbb{I}$ to denote the empty object, which can be viewed as a combinatorial class containing a single element of size 0. For rooted trees, the combinatorial class containing a single vertex $\bullet$ which is an object of size 1 is also very useful.

For example, let $\mathscr{B}$ be a class of trees. Then $\bullet \times \mathscr{B} \times \mathscr{B}$ is the combinatorial class of ordered triples consisting of a single vertex and two trees from $\mathscr{B}$. Viewing the first $\bullet$ as a new root and the two trees from $\mathscr{B}$ as children of the root, we can interpret $\bullet \times \mathscr{B} \times \mathscr{B}$ as the combinatorial class of nonempty rooted trees where the root has two children, each a tree from $\mathscr{B}$, either of which may be empty if $\mathbb{I} \in \mathscr{B}$. For the purposes of an ordinary or exponential generating function, or any other evaluation of the augmented generating function where the evaluation depends only on the size of the objects, there is no difference between the ordered triple of a vertex and two trees on the one hand and the tree with root and those trees as children on the other hand. Consequently we can view

$$\mathscr{B} = \mathbb{I} + \bullet \times \mathscr{B} \times \mathscr{B}$$

as a specification for **binary rooted trees**. These trees are binary in the sense that each vertex has at most two children. More specifically, since the empty tree is allowed, empty children are allowed because the $\mathbb{I}$ in the specification means $\mathbb{I} \in \mathscr{B}$ and so empty is possible in either or both of the $\mathscr{B}$ in the second term. Also all children are designated as left or right even when there is only one child, because $\mathscr{B} \times \mathscr{B}$ gives ordered pairs so having the first empty and the second nonempty is different from having the first nonempty and the second empty. That is

$$\mathscr{B} = \{\mathbb{I}, \bullet, \text{\Large$\nearrow$}, \text{\Large$\searchArrow$}, \text{\Large$\wedge$}, \ldots\}.$$

Another important operation is the sequence operation.

Definition 4 For a combinatorial class $\mathscr{C}$, $\mathrm{Seq}(\mathscr{C})$ is the combinatorial class whose elements are ordered lists (possibly empty) of objects from $\mathscr{C}$ with size the product of the component objects. Equivalently

$$\mathrm{Seq}(\mathscr{C}) = \mathbb{I} + \mathscr{C} + \mathscr{C} \times \mathscr{C} + \mathscr{C} \times \mathscr{C} \times \mathscr{C} + \cdots .$$

For example, the class of plane rooted trees, that is rooted trees where the children of each vertex are ordered, has the specification

$$\mathscr{T} = \bullet \times \mathrm{Seq}(\mathscr{T}).$$

There are many other operations which will not be as useful for us, such as the operation of taking a cycle of objects from a given class, see Theorem I.1 of [2].

Combinatorial operations are useful because they translate into functional equations for the generating function. As given above, specifications don't really keep all the information we want. For example in saying that binary trees are specified by

$$\mathscr{B} = \mathbb{I} + \bullet \times \mathscr{B} \times \mathscr{B}$$

it is left implicit that the $\bullet$ represents the root and the two $\mathscr{B}$ are the left and right subtrees. To write the same specification at the level of the augmented generating function we need notation which is more explicit in this regard.

Definition 5 Let $t_1, \ldots, t_n$ be rooted trees. Then $B_+(t_1 \cdots t_n)$ is the rooted tree which consists of a new root with each of $t_1, \ldots, t_n$ as its children.

For example

$$B_+(\overset{\bullet}{\underset{\bullet}{\bullet}}) = \overset{\bullet\;\bullet}{\underset{\bullet}{\bigwedge}}$$

If we view $t_1 \cdots t_n$ simply as a disjoint union of trees then B_+ returns a tree with no order information on the children of its root. If instead we view $t_1 \cdots t_n$ as the ordered list $(t_1, \ldots, t_n)$ then B_+ returns a tree with ordered children having the same order as in the list.

Then we can rewrite the above specifications as functional equations for the augmented generating function.

$$\mathscr{B} = \mathbb{I} + \bullet \times \mathscr{B} \times \mathscr{B} \quad \leadsto \quad B(x) = \mathbb{I} + x B_+(B(x)^2)$$

and

$$\mathscr{T} = \bullet \times \mathrm{Seq}(\mathscr{T}) \quad \leadsto \quad T(x) = x B_+(\mathrm{Seq}(T(x)))$$

Notice that the sequence operator gives a geometric series in the original class. So we could use the notation

$$\frac{1}{1 - \mathscr{T}}$$

in place of Seq(T). This is quite common for Dyson-Schwinger equations and so we
usually see

$$\mathscr{T} = \bullet \times \text{Seq}(\mathscr{T}) \quad \rightsquigarrow \quad T(x) = x B_+\left(\frac{1}{1 - T(x)}\right)$$

Note that we can interpret an equation like

$$T(x) = \mathbb{I} + x B_+(T(x)^2)$$

in two ways. If we view the argument to B_+ as being an ordered pair of series,
and hence any term in its expansion is an ordered pair of trees, then the result
is the augmented generating function for binary trees with left and right children
distinguished as discussed above.

On the other hand we could also view the argument to B_+ as being unordered.
Then, for instance, the term $\overset{\bullet}{\underset{\bullet}{}}$ appears twice. The result is a series with coefficients
which are linear combinations of rooted trees where the trees themselves have with
no additional structure. In this set up each tree appears with the multiplicity corresponding to the number of binary trees from $\mathscr{B}$ with that underlying shape.

From this viewpoint, starting with

$$T(x) = \mathbb{I} + x B_+(T(x)^2)$$

we would get the augmented generating function

$$T(x) = \mathbb{I} + \bullet x + 2\overset{\bullet}{\underset{\bullet}{}}x^2 + \left(\bigwedge + 4\overset{\bullet}{\underset{\bullet}{\bullet}}\right)x^3 + O(x^4)$$

We will predominantly take this latter viewpoint and so, unless otherwise specified, augmented generating functions of trees will be viewed as having coefficients in
linear combinations of rooted trees where the rooted trees have no order information
at the vertices nor other additional structure.

With this convention in mind, given a formal power series $A(x)$ with nonzero constant term we can write the following functional equation of augmented generating
functions

$$T(x) = x B_+(A(T(x))$$

where $A(T(x))$ is simply the composition of formal power series. Tree classes which
can be built in this way are known as **simple tree classes**. We can extend this by
allowing a family of B_+ operators indexed by a label which is associated to the
new root vertex. Furthermore the labelled roots may not all have size 1. Given a
combinatorial class $\mathscr{L}$ of labels, and a formal power series $A_a(x)$ for each $a \in \mathscr{L}$,
this would give functional equations of the following form

$$T(x) = \sum_{a \in \mathscr{L}} x^{|a|} B_+^a(A_a(T(x))).\tag{3.2}$$

We need to take a moment now to consider conventions with regards to $\mathbb{I}$. The definition of simple tree classes given above does not allow $T(x)$ to begin with a constant term. Instead $A(x)$ must have a nonzero constant term in order to get the recurrence going. There is no loss of generality here because if we wanted some series $\widetilde{T}(x) = c\mathbb{I} + O(x)$ then we simply use $T(x) = \widetilde{T}(x) - c\mathbb{I}$ instead. Having said that, it isn't convenient to always be adjusting the constant term and keeping it in lets us stay closer to the Dyson-Schwinger equations of physics where the constant term corresponds to the tree-level contribution. Therefore, we will allow the labelling class in (3.2) to include an element $0 \in \mathscr{L}$ of size 0 and take the convention that $B_+^0(A_0(T(x)))$ is simply a constant $a_0\mathbb{I}$.

For example we could write the specification

$$B(x) = \mathbb{I} + xB_+(B(x)^2)$$

with $\mathscr{L} = \{0, 1\}$ as

$$B(x) = \sum_{a \in L} x^a B_+^a(B(x)^2)$$

and we could modify the specification

$$T(x) = xB_+\left(\frac{1}{1 - T(x)}\right)$$

to $U(x) = \mathbb{I} - T(x)$ giving

$$U(x) = \mathbb{I} - xB_+\left(\frac{1}{U(x)}\right)$$

where $1/U(x)$ should be interpreted as shorthand for a geometric series expansion (which makes sense since $U(x) = \mathbb{I} + O(x)$). Then, again with $\mathscr{L} = \{0, 1\}$, this fits into the present framework.

We can also form systems of equations in a similar way

$$T_1(x) = \sum_{a \in \mathscr{L}} x^{|a|} B_+^{a,1}(A_{a,1}(T(x)))$$

$$T_2(x) = \sum_{a \in \mathscr{L}} x^{|a|} B_+^{a,2}(A_{a,2}(T(x)))$$

$$\vdots$$

$$T_n(x) = \sum_{a \in \mathscr{L}} x^{|a|} B_+^{a,n}(A_{a,n}(T(x)))$$

(3.3)

From a physics perspective, equations of these form capture the combinatorial form of a Dyson-Schwinger equation. For example if we insert

into itself recursively in all possible ways, we get a rooted tree structure as in the example in Sect. 3.1. We will see more physically relevant Feynman graph examples in Chap. 5. Inspired by this structure, any equation of the form (3.3), either a single equation or a system, with solutions given by augmented generating functions in any renormalization Hopf algebra (such as rooted trees or Feynman graphs, see Chaps. 4 and 5 for more details) will be known as a **combinatorial Dyson-Schwinger equation**.

References

1. Grinberg, D., Reiner, V.: Hopf algebras in combinatorics. arXiv:1409.8356
2. Flajolet, P., Sedgwick, R.: Analytic Combinatorics. Cambridge University Press, Cambridge (2009)
3. Yeats, K.: Rearranging Dyson-Schwinger equations. Mem. Amer. Math. Soc. **211**, 1–82 (2011)
4. Yeats, K.A.: Growth estimates for Dyson-Schwinger equations. Ph.D. thesis, Boston University (2008)
5. Kreimer, D.: On overlapping divergences. Commun. Math. Phys. **204**(3), 669–689 (1999). arXiv:hep-th/9810022
6. Figueroa, H., Gracia-Bondía, J.M.: Combinatorial Hopf algebras in quantum field theory I. Rev. Math. Phys. **17**, 881–976 (2005). arXiv:hep-th/0408145
7. Borinsky, M.: Algebraic lattices in QFT renormalization. Lett. Math. Phys. **106**(7), 879–911 (2016)
8. Schmitt, W.R.: Incidence Hopf algebras. J. Pure Appl. Algebra **96**(3), 299–330 (1994)
9. Knuth, D.: The Art of Computer Programming, vol. 3. Addison-Wesley, Boston, MA (1973)
10. Panzer, E.: Hopf-algebraic renormalization of Kreimer's toy model. Master's thesis, Humboldt-Universität zu Berlin (2011)

Chapter 4
The Connes-Kreimer Hopf Algebra

4.1 Combinatorial Hopf Algebras

If we have some product on combinatorial objects then we would expect this product to take two of the objects and give back an object which in some reasonable sense is the result of combining the original two objects. A very simple example which we have already discussed is disjoint union. Another example is the concatenation of words. Let Ω be an alphabet and let w_1, w_2 be words over the alphabet (that is ordered lists of elements of Ω). Then the concatenation of w_1 and w_2, written $w_1 w_2$, is simply the word made of w_1 immediately followed by w_2.

Other reasonable combinatorial products don't give back a single object but rather a multiset of them. One important example of this is the shuffle product of words. Let Ω again be an alphabet and let w_1 and w_2 be words over Ω. Then, a shuffle of w_1 and w_2 is a word whose letters can be partitioned into two parts so that one part consists of the letters of w_1 in order and the other part consists of the letters of w_2 in order. For example

$$axbcy$$

is a shuffle of abc and xy. Suppose we are interested in all shuffles of two words. Algebraically represent the multiset of these shuffles as a formal sum. Work now in a vector space or module of linear combinations of words and the shuffle gives a product, called the **shuffle product** which we notate with $\shuffle$ and can define recursively by

$$aw_1 \shuffle bw_2 = a(w_1 \shuffle bw_2) + b(aw_1 \shuffle w_2)$$

and

$$\mathbb{I} \shuffle w = w \shuffle \mathbb{I} = w$$

where $a, b \in \Omega$, w_1, w_2 and w are words over Ω, and $\mathbb{I}$ is the empty word.

Dually, we may want to break up combinatorial objects into pieces. A coproduct accomplishes this. It takes an object and returns a sum of ways to break the object

© The Author(s) 2017
K. Yeats, *A Combinatorial Perspective on Quantum Field Theory*,
SpringerBriefs in Mathematical Physics 15, DOI 10.1007/978-3-319-47551-6_4

into two pieces. We get a combinatorial Hopf algebra if the product and coproduct
are compatible in a specific way.

One can jump right in and play with combinatorial Hopf algebras just by working
concretely with these two operations and not worrying much about the algebraic
underpinnings. With this in mind, some readers may want to skip to Sect. 4.2. Ultimately, however, a solid abstract foundation is extremely powerful (which is one of
the central messages of mathematics as a whole), and so next we will briefly go over
the basic definitions for Hopf algebras. A good references on combinatorial Hopf
algebras is [1].

To make the duality between coproducts and products most clear the definitions
for Hopf algebras are best presented using commutative diagrams.

Definition 6 An *algebra* A over K is a vector space over K with two linear maps
$m : A \otimes A \to A$, called the product or multiplication, and $u : K \to A$, called the
unit, such that the following two diagrams

$$
\begin{array}{ccc}
A \otimes A \otimes A & \xrightarrow{\mathrm{id} \otimes m} & A \otimes A \\
\downarrow{\scriptstyle m \otimes \mathrm{id}} & & \downarrow{\scriptstyle m} \\
A \otimes A & \xrightarrow{\quad m \quad} & A
\end{array}
$$

and

$$
\begin{array}{ccccc}
K \otimes A & \xleftarrow{a \mapsto 1 \otimes a} & A & \xrightarrow{a \mapsto a \otimes 1} & A \otimes K \\
\downarrow{\scriptstyle u \otimes \mathrm{id}} & & \downarrow{\scriptstyle \mathrm{id}} & & \downarrow{\scriptstyle \mathrm{id} \otimes u} \\
A \otimes A & \xrightarrow{\quad m \quad} & A & \xleftarrow{\quad m \quad} & A \otimes A
\end{array}
$$

commute.

For those not fluent in commutative diagrams lets check that this corresponds to
what we usually think of as a K-algebra. Elementarily one thinks of the product
in an algebra as a bilinear map from $A \times A$ to A. Converting bilinear maps to
linear maps is exactly what the tensor product does, so this is equivalent to thinking
of the product as a linear map from $A \otimes A$ to A. The remaining property of a
product is associativity. To see associativity, read off the first commutative diagram:
take an elementary tensor $a \otimes b \otimes c \in A \otimes A \otimes A$, then the diagram tells
us that $m(m(a \otimes b) \otimes c) = m(a \otimes m(b \otimes c))$, or in more mundane language
$(a \cdot b) \cdot c = a \cdot (b \cdot c)$.

Usually we think of the unit as distinguished element of A which is a multiplicative
identity. But once we know where 1 is in A, then we also know where $1 + 1$ is and
$1/(1 + 1)$ and so on, so the unit is telling us how to see K inside A, just as the map
u does. The unit in the usual sense is given by $u(1) \in A$. The second commutative
diagram says $u(1) \cdot a = a \cdot u(1) = a$.

A product tells us how to combine two elements together. A coproduct does the opposite, it tells us how to take an element apart. Precisely, we simply reverse all the arrows in the definition of an algebra to get the following definition.

Definition 7 A *coalgebra* C over K is a vector space over K with two linear maps $\Delta : C \to C \otimes C$, called the coproduct, and $\varepsilon : C \to K$, called the counit, such that the following two diagrams

$$
\begin{array}{ccc}
C \otimes C \otimes C & \xleftarrow{\ id \otimes \Delta\ } & C \otimes C \\
\uparrow{\scriptstyle \Delta \otimes id} & & \uparrow{\scriptstyle \Delta} \\
C \otimes C & \xleftarrow{\quad \Delta \quad} & C
\end{array}
$$

and

$$
\begin{array}{ccccc}
K \otimes C & \xrightarrow{\ k \otimes c \mapsto kc\ } & C & \xleftarrow{\ c \otimes k \mapsto kc\ } & C \otimes K \\
\uparrow{\scriptstyle \varepsilon \otimes id} & & \uparrow{\scriptstyle id} & & \uparrow{\scriptstyle id \otimes \varepsilon} \\
C \otimes C & \xleftarrow{\ \Delta\ } & C & \xrightarrow{\ \Delta\ } & C \otimes C
\end{array}
$$

commute.

We will be interested in situations where we have both a product and a coproduct so we need to understand when a product and coproduct are compatible. The answer is that the coproduct and counit need to be algebra homomorphisms or equivalently the product and unit need to be coalgebra homomorphisms. To make this precise we need the definition of an algebra homomorphism phrased in this language:

Definition 8 Let A and B be algebras over K. Then $f : A \to B$ is an *algebra homomorphism* if the following diagrams commute

$$
\begin{array}{ccc}
A & \xrightarrow{\ f\ } & B \\
\uparrow{\scriptstyle m_A} & & \uparrow{\scriptstyle m_B} \\
A \otimes A & \xrightarrow{\ f \otimes f\ } & B \otimes B
\end{array}
$$

$$
\begin{array}{ccc}
 & K & \\
{\scriptstyle u_A}\swarrow & & \searrow{\scriptstyle u_B} \\
A & \xrightarrow{\ f\ } & B
\end{array}
$$

A *coalgebra homomorphism* is the equivalent definition with arrows reversed. It is a good exercise for the reader to check that given a vector space which is simultaneously an algebra and a coalgebra over K, saying that the product and unit are coalgebra homomorphisms is equivalent to saying that the coproduct and counit are algebra homomorphisms (the relevant commutative diagrams end up just being the same). This is the compatibility we want.

Definition 9 Suppose B is both an algebra and a coalgebra over K and that the coproduct and counit are algebra homomorphisms. Then, B is a ***bialgebra*** over K.

The most important bialgebras for us will be the Connes-Kreimer Hopf algebra which will be defined in Sect. 4.2 and renormalization Hopf algebras of Feynman graphs which will be defined in Chap. 5. Two simpler, but still useful, examples can be made from words as follows.

Let $\mathscr{W}$ be the set of all words over an alphabet Ω. Let $W = \text{span}_K(\mathscr{W})$ be the vector space of formal linear combinations of words. We can make W into a bialgebra in two ways. In both cases the unit map will be $1 \mapsto \mathbb{I}$, with $\mathbb{I}$ being the empty word, and the counit will be $\varepsilon(\mathbb{I}) = 1$, $\varepsilon(w) = 0$ for a nonempty word, both extended as homomorphisms.

To make W into the ***shuffle-deconcatenation*** bialgebra, let the product be the shuffle, ⊔⊔, and let the coproduct be given by

$$\Delta(a_1 a_2 \cdots a_n) = \sum_{i=0}^{n} a_1 \cdots a_i \otimes a_{i+1} \cdots a_n$$

for a nonempty word $a_1 a_2 \cdots a_n$ and extended as an algebra homomorphism.

To make W into the ***concatenation-deshuffle*** bialgebra, let the product be concatenation and let the coproduct be given by

$$\Delta(a_1 a_2 \cdots a_n) = \sum_{I \subseteq \{1,\dots,n\}} a_I \otimes a_{\{1,\dots,n\} \setminus I}$$

where a_I denotes the subword of $a_1 a_2 \cdots a_n$ consisting of the letters indexed by I.

Here are some useful definitions and results; for more details see [1, Sect. 1.3].

An algebra A is ***commutative*** if the multiplication is commutative, that is $m(a \otimes b) = m(b \otimes a)$. If we let $\tau : a \otimes b \mapsto b \otimes a$ be the transposition operation then we can write commutativity as the following commutative diagram

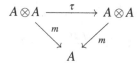

A coalgebra is ***cocommutative*** if the reverse diagram holds. Concretely, this means that one can flip all the tensors in the coproduct and get the same result. The shuffle-deconcatentation bialgebra is commutative but not cocommutative while the concatenation-deshuffle bialgebra is cocommutative but not commutative.

A vector space V over K is ***graded*** (or $\mathbb{Z}_{\geq 0}$-graded to be more precise) if it has a direct sum decomposition $V = \bigoplus_{n=0}^{\infty} V_n$. The vector space V_n is called the ***graded piece*** of degree n and the elements of V_n are called ***homogeneous*** of degree n. A linear map $f : V \to W$ between graded vector spaces is itself ***graded*** if $f(V_n) \subseteq W_n$ for all n. Note that if V is a graded vector space, then $V \otimes V$ is also graded. Specifically, the graded piece of degree n in $V \otimes V$ is $\bigoplus_{j=0}^{n} V_j \otimes V_{n-j}$. An algebra, coalgebra, or bialgebra is **graded** if the underlying vector space and all the defining maps are graded.

For example, both word bialgebras we have defined are graded. If we have any combinatorial class $\mathscr{C}$ then we can form a graded vector space $\mathrm{span}(\mathscr{C}) = \bigoplus_{n=0}^{\infty}\mathrm{span}(\mathscr{C}_n)$. Then if we have a compatible "putting together" map for multiplication and "taking apart" map for comultiplication we get a combinatorial bialgebra (and we'll soon see it is in fact a combinatorial Hopf algebra). The combinatorial bialgebras we'll be working with are slightly more special because we'll be using disjoint union as our multiplication, so if $\mathscr{C}$ is a combinatorial class of connected objects, then $K[\mathscr{C}]$ is a graded vector space where the homogeneous elements of degree n are the homogeneous polynomials of degree n (with each element $c \in \mathscr{C}$ being of degree equal to its size). To form a combinatorial bialgebra in this more restricted context we only need to find a compatible coproduct.

A graded vector space V over K is **connected** if $V_0 \cong K$. The two word bialgebras are both connected. In fact, in our combinatorial context the connected condition is very natural because often for a combinatorial class $\mathscr{C}$ we have $\mathscr{C}_0 = \{\mathbb{I}\}$, that is we have a single element of size 0 which we call the empty object. In this case the degree 0 graded piece is $\mathrm{span}(C_0) = K\mathbb{I} \cong K$.

Here are a few simple facts about graded connected bialgebras. The proofs are straightforward and good exercises.

Proposition 1 *Let A be a graded connected bialgebra over K.*

1. *$u : K \to A_0$ is an isomorphism.*
2. *$\varepsilon|_{A_0} : A_0 \to K$ is the inverse isomorphism to u.*
3. *$\ker \varepsilon = \bigoplus_{n=0}^{\infty} A_n$.*
4. *For $x \in \ker \varepsilon$, $\varDelta(x) = \mathbb{I} \otimes x + x \otimes \mathbb{I} + \widetilde{\varDelta}(x)$ where $\widetilde{\varDelta}(x) \in \ker \varepsilon \otimes \ker \varepsilon$.*

For a bialgebra A and $a \in A$, if $\varDelta(a) = a \otimes a$ then we say a is **group-like**. Group-like elements will not be very important for us since in a graded connected bialgebra the only one is $\mathbb{I}$, however series of elements in our Hopf algebras can be group-like. For a bialgebra A and $a \in A$, if $\varDelta(a) = \mathbb{I} \otimes a + a \otimes \mathbb{I}$ then we say a is **primitive**. If $\varDelta(a) = \mathbb{I} \otimes a + a \otimes \mathbb{I} + \widetilde{\varDelta}(a)$, then the $\mathbb{I} \otimes a + a \otimes \mathbb{I}$ is the **primitive part**.

For us the most important thing about graded connected bialgebras is that we get for free that they are not just bialgebras but actually Hopf algebras. This means that we have one more map, called the antipode. To understand it we first need to understand the convolution product.

Definition 10 Let C be a coalgebra and A an algebra. Let $f, g : C \to A$ be linear maps. The **convolution product** of f and g is

$$f \star g = m \circ (f \otimes g) \circ \varDelta.$$

Definition 11 A bialgebra A is a **Hopf algebra** if there exists a linear map $S : A \to A$, called the **antipode**, satisfying the following commutative diagram

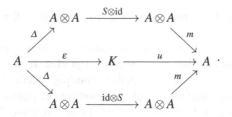

Equivalently, interpreting this as a formula, the defining property of the antipode is $S \star \mathrm{id} = \mathrm{id} \star S = u\varepsilon$.

Here are a few important properties of the antipode (see Sect. 1.4 of [1] for more details). The last of these is the one which is key for us; it says that the antipode comes for free in our context.

Proposition 2 *1. Let A be a Hopf algebra. The antipode S is an algebra anti-automorphism. That is, $S(\mathbb{I}) = \mathbb{I}$ and $S(ab) = S(b)S(a)$*
2. Let A be a Hopf algebra. If A is commutative or cocommutative then $S \circ S = \mathrm{id}$.
3. Let A be a graded connected bialgebra. A has a unique antipode S which is determined recursively. Furthermore, S is a graded map, so A is a graded Hopf algebra.

For the first two items, see [1, Sect. 1.4] for proofs. The last point is particularly important for us, so let's go through the calculation

Proof (Proof of 3.) Begin with $S \star \mathrm{id} = u\varepsilon$ and turn this into a recurrence. We have

$$A = \bigoplus_{n=0}^{\infty} A_n$$

and $A_0 = K$. By the first point of the proposition $S(\mathbb{I}) = \mathbb{I}$ so $S|_{A_0} = \mathrm{id}$. For x of homogeneous degree $n > 0$ by the previous proposition we can write $\Delta(x) = \mathbb{I} \otimes x + x \otimes \mathbb{I} + \tilde{\Delta}(x)$ with $\tilde{\Delta}(x) \in \ker \varepsilon \otimes \ker \varepsilon$. Write

$$\tilde{\Delta}(x) = \sum_i x_{i,1} \otimes x_{i,2}$$

then each $x_{i,j}$ has degree strictly less than n. Then

$$0 = u\varepsilon(x) = (S \star \mathrm{id})(x) = x + S(x) + \sum_i S(x_{i,1})x_{i,2}$$

so

$$S(x) = -x - \sum_i S(x_{i,1})x_{i,2}$$

which determines S recursively.

Following through this definition for the concatenation-deshuffle Hopf algebra we get $S(a_1 a_2 \ldots a_n) = (-1)^n a_n \ldots a_2 a_1$. The word is reversed and there is a sign.

There are a few things to note here. First, the antipode is the direct analogue to the Möbius function in this context. This explains why the recursive formula is strongly reminiscent of Möbius inversion. See [2] for more about the connection between the antipode and Möbius inversion. Second, since the combinatorial Hopf algebras of interest to us are commutative and hence $S \circ S = \mathrm{id}$, these Hopf algebras are not the kind of interest in the quantum groups world.

Note also that augmented generating functions fit very nicely into this algebraic context because the coefficients are simply elements of the relevant combinatorial Hopf algebra H. Algebraic operations can be extended from H to $H[[x]]$ and can also be used to speak about the generating function.

4.2 The Connes-Kreimer Hopf Algebra of Rooted Trees

The most important combinatorial Hopf algebra for us is the **Connes-Kreimer Hopf algebra of rooted trees**. It is what we need to capture the structure of renormalization (see Sect. 4.3) and is algebraically important (see Sect. 4.4).

Let $\mathscr{T}$ be the combinatorial class of rooted trees with no plane structure and without including the empty tree. Let $\mathscr{H} = K[\mathscr{T}]$. As in Sect. 3.1 think of $\mathscr{H}$ as a space of forests. The empty forest $\mathbb{I}$ reappears as the empty monomial. The algebra structure of $\mathscr{H}$ is the algebra structure we want for the Connes-Kreimer Hopf algebra. Recall, given $t \in \mathscr{T}$ and $v \in V(T)$, t_v is the subtree of t rooted at v (see Sect. 3.1). The **coproduct**, Δ, is defined as follows: for $t \in \mathscr{T}$

$$\Delta(t) = \sum_{\substack{C \subseteq V(t) \\ C \text{ antichain}}} \left(\prod_{v \in C} t_v \right) \otimes \left(t - \prod_{v \in C} t_v \right)$$

and Δ is extended to $\mathscr{H}$ as an algebra homomorphism.

For example

$$\Delta\left(\vcenter{\hbox{\includegraphics{tree}}} \right) = \vcenter{\hbox{\includegraphics{tree}}} \otimes \mathbb{I} + \mathbb{I} \otimes \vcenter{\hbox{\includegraphics{tree}}} + {}^\bullet \otimes {\textstyle\vcenter{\hbox{\includegraphics{tree}}}} + {}^\bullet \otimes \vcenter{\hbox{\includegraphics{tree}}} + {}^{\bullet\bullet} \otimes {\textstyle\vcenter{\hbox{\includegraphics{tree}}}} + {\textstyle\vcenter{\hbox{\includegraphics{tree}}}} \otimes {\textstyle\vcenter{\hbox{\includegraphics{tree}}}} + {}^{\bullet\bullet}{\textstyle\vcenter{\hbox{\includegraphics{tree}}}} \otimes {}^\bullet.$$

For a forest example, using multiplicativity we have

$$\Delta\left(\overset{\bullet\bullet}{\underset{\bullet}{\mathsf{I}}}\right) = \left(\overset{\bullet}{\underset{\bullet}{\mathsf{I}}}\otimes\mathbb{I} + {}^{\bullet}\otimes{}^{\bullet} + \mathbb{I}\otimes\overset{\bullet}{\underset{\bullet}{\mathsf{I}}}\right)\left({}^{\bullet}\otimes\mathbb{I} + \mathbb{I}\otimes{}^{\bullet}\right)$$
$$= \overset{\bullet\bullet}{\underset{\bullet}{\mathsf{I}}}\otimes\mathbb{I} + {}^{\bullet\bullet}\otimes{}^{\bullet} + {}^{\bullet}\otimes\overset{\bullet}{\underset{\bullet}{\mathsf{I}}} + \overset{\bullet}{\underset{\bullet}{\mathsf{I}}}\otimes{}^{\bullet} + {}^{\bullet}\otimes{}^{\bullet\bullet} + \mathbb{I}\otimes\overset{\bullet\bullet}{\underset{\bullet}{\mathsf{I}}}.$$

Another way to think of Δ is in terms of sets of edges to cut at rather than sets of vertices to root at. For any antichain $C \subseteq V(T)$ which does not contain the root (and hence is not the singleton of the root alone), take the edges immediately above the elements of C. This set of edges has the property that no two are on the same path from a leaf to the root and every set of edges with this property comes from an antichain of vertices. If we think of cutting these edges, then the resulting subtrees which do not contain the original root are precisely $\prod_{v\in C} t_v$, while the unique subtree containing the original root is $t - \prod_{v\in C} t_v$. The antichain consisting of the root alone is a special case, giving the summand $t \otimes \mathbb{I}$. In edge-cut language we think of this as a virtual cut called the **empty cut** which we can visualize as cutting above the root in order to detach the entire tree.

The counit of the Connes-Kreimer Hopf algebra is fairly uninteresting. It is the algebra homomorphism $\varepsilon : \mathscr{H} \to K$ which takes any (nonempty) $t \in \mathscr{T}$ to 0 and takes $\mathbb{I}$ to 1.

Δ and ε make $\mathscr{H}$ into a bialgebra. The required properties are mostly very easy to check. The only nontrivial one is coassociativity. Even still the idea is not difficult— taking an antichain of vertices of a tree and then taking another antichain which lies beneath it in the poset in all possible ways is the same as taking an antichain of vertices and then taking another which lies above it in all possible ways. In either case you are simply taking two antichains, one above the other, in all possible ways.

$\mathscr{H}$ is graded by the number of vertices of a forest. The degree zero piece is $K\mathbb{I}$ so $\mathscr{H}$ is graded and connected. Thus by the results of the previous section $\mathscr{H}$ has an antipode and so is a Hopf algebra. Concretely, this means the antipode is given by the following formula

$$S(t) = -t - \sum_{\substack{\varnothing\subsetneq C\subsetneq V(t)\\ C\text{ antichain}}} S\left(\prod_{v\in C} t_v\right)\left(t - \prod_{v\in C} t_v\right)$$

for $t \in \mathscr{T}$.

For example

$$S(\bullet) = -\bullet$$
$$S(\bullet\bullet) = -\bullet\bullet - 2S(\bullet)\bullet = \bullet\bullet$$
$$S(\overset{}{\underset{\bullet\ \bullet}{\wedge}}) = -\overset{}{\underset{\bullet\ \bullet}{\wedge}} - 2S(\bullet)\overset{\bullet}{\underset{\bullet}{\mathsf{I}}} - S(\bullet\bullet)\bullet = -\overset{}{\underset{\bullet\ \bullet}{\wedge}} + 2\bullet\overset{\bullet}{\underset{\bullet}{\mathsf{I}}} - \bullet\bullet\bullet$$

The antipode will be useful because it captures the recursive structure of renormalization as we will see in the next section.

4.3 Physical Properties

The Connes-Kreimer Hopf algebra of rooted trees was introduced in order to give an algebraic underpinning to the BPHZ renormalization prescription (see below). Let's step back a moment and see where renormalization comes from. In perturbative quantum field theory we have series expansions indexed by Feynman diagrams. Each Feynman diagram contributes an integral but these integrals, in key cases, are divergent. One way to think of this intuitively is that the quantities in question don't make sense absolutely but do make sense relatively. This is meant in the sense that the integrals diverge, but if we take the difference, formally, between the integral at some fixed reference scale and the integral at some other scale of interest then we get a finite integral. Thinking this way leads, after all the substantial details are worked out, to renormalization by subtracting at a reference scale such as in MOM scheme. See below for some toy tree Feynman rules which can be renormalized in this way. There are many other ways to think about the divergences in quantum field theory and many other renormalization schemes with different strengths and weaknesses.

One of these substantial details is of key combinatorial importance. Specifically it is not necessarily just the overall integral which is divergent, but often partial integrations are already divergent. This corresponds to subgraphs of the Feynman graph being divergent. Fortunately this problem can be dealt with using a recursive subtraction scheme. This approach was worked out over the course of roughly a decade. Bogoliubov and Parasiuk took one of the key steps with a tool that came to be known as Bogoliubov's $\overline{R}$-map [3]. Overlapping subdivergences are particularly tricky and Zimmerman, thinking in terms of trees of subdivergences within graphs, gave what is known as the Zimmerman forest formula [4] to understand their renormalization. In the end the recursive renormalization technique which was developed is known as BPHZ renormalization after Bogoliubov, Parasiuk, Hepp, and Zimmermann, see [5] for an overview.

Much more recently Kreimer [6, 7] realized the underlying structure of BPHZ renormalization is captured by a combinatorial Hopf algebra. This gave a new and insightful underpinning to renormalization and opened the door to interactions with a variety of different parts of mathematics. These parts of mathematics had new contributions to make to quantum field theory and new things to learn from quantum field theory, for example [8–29]. I count myself as such a mathematician on the combinatorial side.

In this Hopf algebraic framework, the antipode is capturing the structure of recursive renormalization, and after twisting the antipode with the Feynman rules in an appropriate way we can simply write down the renormalized map in this algebraic language. In Chap. 5 we'll develop Feynman graphs and Feynman rules as well as renormalization Hopf algebras directly at the level of graphs. For now, however, we will stick to rooted trees. Because of Theorem 1 (see Sect. 4.4) we lose very little by working at the level of trees, though working directly with the Feynman graph Hopf algebras can sometimes be more appealing.

In order to be more concrete let's make some toy Feynman rules for trees which are more realistic than the tree factorial Feynman rules and in particular require renormalization. Then we'll use this to illustrate how renormalization works. This toy example can be found in [19] and a nice exposition of it is in [30]—more details on it can be found there.

Recursively define a map ϕ_s on the Connes-Kreimer Hopf algebra of rooted trees, $\mathcal{H}$, as follows. Require ϕ_s to be an algebra homomorphism, then for any nonempty forest f

$$\phi_s(B_+(f)) = \int_0^\infty \frac{\phi_z(f)}{s+z} dz$$

where s is a parameter with $s > 0$. This ϕ_s is the Feynman rules for this toy example. The recursive appearance of ϕ inside the definition has a different argument because the integrals themselves are nested as they are for real quantum field theories.

Let's see what happens in a few simple examples.

$$\phi_s(\bullet) = \int_0^\infty \frac{1}{s+z} dz$$

This integral is already divergent since the antiderivative of $\frac{1}{s+z}$ is $\log(s+z)$ which diverges as $z \to \infty$. This means that we should think of the target space of ϕ_s as being a space of formal integral expressions. However, we are interested in doing something to these formal integral expressions in order to get expressions which are integrable.

In this case we can renormalize by subtraction. Take the difference of the integrands of $\phi_s(\bullet)$ and $\phi_1(\bullet)$ giving

$$\frac{1}{s+z} - \frac{1}{1+z} = \frac{1-s}{(s+z)(1+z)}$$

The s parameter is acting like a kinematical parameter and taking this difference means looking at our quantity relatively not absolutely. This difference is integrable for $0 \leq z \leq \infty$:

$$\int_0^\infty \frac{1-s}{(s+z)(1+z)} = -\log(s)$$

and so, being sloppy in the conventional way, we write

$$\phi_s(\bullet) - \phi_1(\bullet) = -\log(s)$$

and we have renormalized $\phi_s(\bullet)$.

Now consider

$$\phi_s\!\left(\overset{\bullet}{\underset{\bullet}{|}}\right) = \int_0^\infty \frac{\phi_{z_1}(\bullet)}{s+z_1}\,dz_1 = \int_0^\infty\int_0^\infty \frac{1}{(z_1+z_2)(s+z_1)}\,dz_2dz_1.$$

This is again divergent but so is $\phi_s\!\left(\overset{\bullet}{\underset{\bullet}{|}}\right) - \phi_1\!\left(\overset{\bullet}{\underset{\bullet}{|}}\right)$. The problem is the inner dz_2 integration—first we need to subtract to take care of it. The way to do this systematically is to use the antipode. Let R be the map which takes a formal integral expression in the parameter s and evaluates it at $s = 1$. Then define

$$S_R^{\phi_s}(\mathbb{I}) = 1$$

$$S_R^{\phi_s}(t) = -R(\phi_s(t)) - \sum_{\substack{\varnothing\subsetneq C\subsetneq V(t)\\ C\,\text{antichain}}} S_R^{\phi_s}\!\left(\prod_{v\in C} t_v\right) R\!\left(\phi_s\!\left(t - \prod_{v\in C} t_v\right)\right)$$

for tree t and extend as an algebra homomorphism to $\mathscr{H}$. We think of $S_R^{\phi_s}$ as a twisted antipode—the defining recurrence says $S_R^{\phi_s} \star R\phi_s = u\varepsilon$ (recall the convolution product of Sect. 4.1). Then the **renormalized Feynman rules** are

$$\phi_{\text{renormalized}} = S_R^{\phi_s} \star \phi_s.$$

Let's calculate for $\overset{\bullet}{\underset{\bullet}{|}}$ to see that it works.

$$S_R^{\phi_s}\!\left(\overset{\bullet}{\underset{\bullet}{|}}\right) = -R(\phi_s\!\left(\overset{\bullet}{\underset{\bullet}{|}}\right)) - S_R^{\phi_s}(\bullet)R(\bullet)$$

$$= -\int_0^\infty\int_0^\infty \frac{1}{(z_1+z_2)(1+z_1)}\,dz_2dz_1 + \int_0^\infty \frac{1}{1+z_2}\,dz_2 \int_0^\infty \frac{1}{1+z_1}\,dz_1$$

$$= -\int_0^\infty\int_0^\infty \frac{1-z_1}{(z_1+z_2)(1+z_1)(1+z_2)}\,dz_2dz_1.$$

This is the **counterterm**. Then

$$\phi_{\text{renormalized}}\left(\overset{\bullet}{\underset{\bullet}{\mid}}\right) = \phi_s\left(\overset{\bullet}{\underset{\bullet}{\mid}}\right) + S_R^{\phi_s}(\bullet)\phi_s(\bullet) + S_R^{\phi_s}\left(\overset{\bullet}{\underset{\bullet}{\mid}}\right)$$

$$= \int_0^\infty \int_0^\infty \frac{1}{(z_1+z_2)(s+z_1)}dz_2 dz_1 - \int_0^\infty \frac{1}{1+z_2}dz_2 \int_0^\infty \frac{1}{s+z_1}dz_1$$

$$- \int_0^\infty \int_0^\infty \frac{1-z_1}{(z_1+z_2)(1+z_1)(1+z_2)}dz_2 dz_1$$

$$= \int_0^\infty \int_0^\infty \frac{1-s}{(z_1+z_2)(1+z_2)(s+z_1)} - \frac{1-z_1}{(z_1+z_2)(1+z_1)(1+z_2)}dz_2 dz_1$$

$$= \int_0^\infty \int_0^\infty \frac{(1-z_1)(1-s)}{(z_1+z_2)(1+z_2)(s+z_1)(1+z_1)}$$

$$= \frac{1}{2}\log^2(c)$$

which is finite.

This whole story works provided R is a Rota-Baxter map, see [5, 31], and furthermore it can be encased in the more general geometric framework of Birkhoff decomposition, see [32, 33] and explained in detail on this particular example in [30].

4.4   Abstract Properties

As well as being physically important, the Connes-Kreimer Hopf algebra, $\mathcal{H}$, is important algebraically because of a universality property discussed below. We can also now characterize what we want algebraically from Feynman rules.

The first step is to capture the nature of B_+ algebraically. Specifically how does B_+ interact with Δ? To answer this question we just check directly from the definition that

$$\Delta(B_+(t)) = (\text{id} \otimes B_+)\Delta(t) + B_+(t) \otimes \mathbb{I}$$

because a cut either cuts off the whole tree or consists of cuts in the subtrees which are the children of the root. It turns out that this identity is telling us that B_+ is a Hochschild 1-cocycle. For readers who do not work with cohomology regularly here is a 3 point summary of how cohomology works:

1. You need a family of maps b_n from objects of size n to objects of size $n+1$ with $b^2 = 0$ (where b^2 means $b_{n+1}b_n$).
2. Take quotients $\ker(b)/\text{im}(b)$.
3. Use these quotients to understand your original objects.

For us we want the objects of size n to be maps from B to $B^{\otimes n}$ for some bialgebra B (most important is the case $B = \mathcal{H}$) Then b is the following map. For any $L: B \to B^{\otimes n}$,

$$bL = (\mathrm{id} \otimes L)\varDelta + \sum_{i=1}^{n}(-1)^i \varDelta_i L + (-1)^{n+1}L \otimes \mathbb{I}$$

where $\varDelta_i = \mathrm{id} \otimes \cdots \otimes \mathrm{id} \otimes \underset{i^{th}\ \text{slot}}{\varDelta} \otimes \mathrm{id} \otimes \cdots \otimes \mathrm{id}$. This gives the Hochschild cohomology of bialgebras.

The first thing you would want to know, following the 3 point summary of cohomology, would be $\ker(b_1)$

$$0 = b_1(L) = (\mathrm{id} \otimes L)\varDelta - \varDelta L + L \otimes \mathbb{I}$$

so $\varDelta L = L \otimes \mathbb{I} + (\mathrm{id} \otimes L)\varDelta$. This is the property B_+ has; it is the property of being 1-cocycle, see [19] or [11] for details.

The pair of $\mathcal{H}$ and B_+ is universal for commutative bialgebras with a 1-cocycle in the following sense.

Theorem 1 *Let A be a commutative algebra and $L : A \rightarrow A$ a map. Then there exists a unique algebra homomorphism $\rho_L : \mathcal{H} \rightarrow A$ such that $\rho_L \circ B_+ = L \circ \rho_L$. If further A is a bialgebra and L is a 1-cocycle then ρ_L is a bialgebra homomorphism and if A is even further a Hopf algebra then ρ_L is a Hopf algebra homomorphism.*

This result is due to Connes and Kreimer [19]. For a nice exposition see Theorem 2.6.4 of [30].

There are two main ways that this theorem tends to be useful. First, take A to be another commutative Hopf algebra with a 1-cocycle. Then by universality we can always map $\mathcal{H}$ with B_+ to A and often we can use this to do the work we need to do in $\mathcal{H}$ instead of A.

Second we can think of A as the target algebra for our Feynman rules. Then any endomorphism of A (playing the role of L in the theorem) induces a ρ which can serve as Feynman rules (see [30, Sect. 3.1]). This now gets to the question of what properties we want Feynman rules to have. There are layers of increasingly restrictive properties we might want to impose.

To begin with, thinking of Feynman rules on graphs rather than rooted trees (see Chap. 5) Feynman rules should be multiplicative on disjoint unions and also have a product property for bridge edges.[1] This multiplicativity over bridges is what lets us move from all Feynman diagrams to one particle irreducible (1PI) diagrams (see Chap. 5). This is done by a Legendre transform, a comprehensive combinatorial treatment of which should appear in upcoming work of Jackson, Kempf, and Morales, see also [34, 35]. Aluffi and Marcolli take these properties as their definition of algebro-geometric Feynman rules in [36]—their goals include finding other examples of such maps which are natural in the context of algebraic geometry.

Treating rooted trees as insertion trees for Feynman graphs, we don't see the reduction to 1PI; we generally consider it to already be done. We still do want maps

---

[1] A *bridge* is an edge which upon removal increases the number of connected components of a graph.

which are multiplicative. In fact they should preserve the algebra structure and so we are led to define Feynman rules as characters, that is algebra homomorphisms from $\mathcal{H}$ to some commutative algebra A. This is the definition from [32].

We have still only touched the surface of the structure of the actual Feynman rules of quantum field theory, so depending on the context we may want to assume more in order to progress. One important additional condition is that the Feynman rules play nicely with B_+. This is necessary to make Dyson-Schwinger equations work. A good algebraic way to impose such a condition is to require that Feynman rules come from an automorphism of a commutative algebra A via Theorem 1. This is what is done in [30, Sect. 3.1].

Alternately one might look to see how the Feynman rules should interact with the coproduct. Restrict to the case where the Feynman rules take values in some ring of polynomials in a variable L. This L will be the L which comes up in the second part of this brief, namely the log of an energy scale. Then the property one would want of Feynman rules is as follows. Let $\phi : \mathcal{H} \to R[L]$ be the Feynman rules. Write $\phi(L_1 + L_2)$ for the map ϕ followed by substituting $L_1 + L_2$ for L. Then the property we want of the Feynman rules is[2]

$$\phi(L_1 + L_2) = \phi(L_1) \star \phi(L_2).$$

These last two properties are closely related as both are essentially telling us that the Feynman rules come from the exponential map on the associated Lie algebra.[3]

References

1. Grinberg, D., Reiner, V.: Hopf algebras in combinatorics. arXiv:1409.8356
2. Schmitt, W.R.: Incidence Hopf algebras. J. Pure Appl. Algebra **96**(3), 299–330 (1994)
3. Bogoliubov, N.N., Parasiuk, O.S.: On the multiplication of causal functions in the quantum theory of fields. Acta Math. **97**, 227–266 (1957)
4. Zimmermann, W.: Convergence of Bogoliubovs method of renormalization in momentum space. Commun. Math. Phys. **15**, 208–234 (1969)
5. Ebrahimi-Fard, K., Kreimer, D.: Hopf algebra approach to Feynman diagram calculations. J. Phys. A **38**, R285–R406 (2005). arXiv:hep-th/0510202
6. Kreimer, D.: On the Hopf algebra structure of perturbative quantum field theories. Adv. Theor. Math. Phys. **2**(2), 303–334 (1998). arXiv:q-alg/9707029
7. Kreimer, D.: On overlapping divergences. Commun. Math. Phys. **204**(3), 669–689 (1999). arXiv:hep-th/9810022
8. Aluffi, P., Marcolli, M.: Parametric Feynman integrals and determinant hypersurfaces. Adv. Theor. Math. Phys. **14**(3), 911–964 (2010). arXiv:0901.2107

---

[2]Personal communication with Spencer Bloch and Dirk Kreimer.

[3]Some more details on the connection between Feynman rules coming from the universal property and the exponential map can be found in lecture notes of Erik Panzer http://people.math.sfu.ca/~kyeats/seminars/Panzer0-02.pdf; my understanding of the connection between the convolution property and the exponential map is based on personal communication with Jason Bell and Julian Rosen.

9. van Baalen, G., Kreimer, D., Uminsky, D., Yeats, K.: The QED beta-function from global solutions to Dyson-Schwinger equations. Ann. Phys. **234**(1), 205–219 (2008). arXiv:0805.0826
10. Bellon, M.: An efficient method for the solution of Schwinger-Dyson equations for propagators. Lett. Math. Phys. **94**(1), 77–86 (2010). arXiv:1005.0196
11. Bergbauer, C., Kreimer, D.: Hopf algebras in renormalization theory: locality and Dyson-Schwinger equations from Hochschild cohomology. IRMA Lect. Math. Theor. Phys. **10**, 133–164 (2006). arXiv:hep-th/0506190
12. Black, S., Crump, I., DeVos, M., Yeats, K.: Forbidden minors for graphs with no first obstruction to parametric feynman integration. Discrete Math. **338**, 9–35 (2015). arXiv:1310.5788
13. Bloch, S., Esnault, H., Kreimer, D.: On motives associated to graph polynomials. Commun. Math. Phys. **267**, 181–225 (2006). arXiv:math/0510011v1 [math.AG]
14. Bloch, S., Kreimer, D.: Feynman amplitudes and Landau singularities for 1-loop graphs. arXiv:1007.0338
15. Broadhurst, D., Schnetz, O.: Algebraic geometry informs perturbative quantum field theory. In: PoS, vol. LL2014, p. 078 (2014). arXiv:1409.5570
16. Brown, F.: On the periods of some Feynman integrals. arXiv:0910.0114
17. Brown, F., Schnetz, O.: Modular forms in quantum field theory. Commun. Number Theor. Phys. **7**(2), 293–325 (2013). arXiv:1304.5342
18. Brown, F., Schnetz, O.: Single-valued multiple polylogarithms and a proof of the zigzag conjecture. J. Number Theory **148**, 478–506 (2015). arXiv:1208.1890
19. Connes, A., Kreimer, D.: Hopf algebras, renormalization and noncommutative geometry. Commun. Math. Phys. **199**, 203–242 (1998). arXiv:hep-th/9808042
20. Foissy, L.: Faà di Bruno subalgebras of the Hopf algebra of planar trees from combinatorial Dyson-Schwinger equations. Adv. Math. **218**(1), 136–162 (2007). arXiv:0707.1204
21. Ebrahimi-Fard, K., Gracia-Bondia, J.M., Patras, F.: A Lie theoretic approach to renormalization. Commun. Math. Phys. **276**(2), 519–549 (2007). arXiv:hep-th/0609035
22. Kreimer, D.: The residues of quantum field theory-numbers we should know. In: Consani, C., Marcolli, M. (eds.) Noncommutative Geometry and Number Theory, pp. 187–204. Vieweg (2006). arXiv:hep-th/0404090
23. Kreimer, D.: A remark on quantum gravity. Ann. Phys. **323**, 49–60 (2008). arXiv:0705.3897
24. Kreimer, D., Sars, M., van Suijlekom, W.: Quantization of gauge fields, graph polynomials and graph cohomology. arXiv:1208.6477
25. Kremnizer, K., Szczesny, M.: Feynman graphs, rooted trees, and Ringel-Hall algebras. Commun. Math. Phys. **289**(2), 561–577 (2009). arXiv:0806.1179
26. Marie, N., Yeats, K.: A chord diagram expansion coming from some Dyson-Schwinger equations. Commun. Number Theory Phys. **7**(2), 251–291 (2013). arXiv:1210.5457
27. Schnetz, O.: Quantum periods: a census of ϕ^4-transcendentals. Commun. Number Theory Phys. **4**(1), 1–48 (2010). arXiv:0801.2856
28. van Suijlekom, W.D.: Renormalization of gauge fields: a Hopf algebra approach. Commun. Math. Phys. **276**, 773–798 (2007). arXiv:0610137
29. Yeats, K.: A few c_2 invariants of circulant graphs. Commun. Number Theor. Phys. **10**(1), 63–86 (2016). arXiv:1507.06974
30. Panzer, E.: Hopf-algebraic renormalization of Kreimer's toy model. Master's thesis, Humboldt-Universität zu Berlin (2011)
31. Ebrahimi-Fard, K., Guo, L., Kreimer, D.: Integrable renormalization ii: the general case. Ann. Henri Poincare **6**, 369–395 (2005). arXiv:hep-th/0403118v1
32. Connes, A., Kreimer, D.: Renormalization in quantum field theory and the Riemann-Hilbert problem I: the Hopf algebra structure of graphs and the main theorem. Commun. Math. Phys. **210**(1), 249–273 (1999). arXiv:hep-th/9912092
33. Connes, A., Kreimer, D.: Renormalization in quantum field theory and the Riemann-Hilbert problem. II: the beta-function, diffeomorphisms and the renormalization group. Commun. Math. Phys. **216**, 215–241 (2001). arXiv:hep-th/0003188
34. Kempf, A., Jackson, D.M., Morales, A.H.: New Dirac delta function based methods with applications to perturbative expansions in quantum field theory. J. Physi. A Math. Theor. **47**(41), 415204 (2014). arXiv:1404.0747

35. Kempf, A., Jackson, D.M., Morales, A.H.: How to (path-) integrate by differentiating. J. Phys. Conf. Ser. **626**, 012,015 (2015). arXiv:1507.04348
36. Aluffi, P., Marcolli, M.: Algebro-geometric Feynman rules. Int. J. Geom. Methods Mod. Phys. **8**, 203–237 (2011). arXiv:0811.2514

Chapter 5
Feynman Graphs

5.1 Half Edge Graphs

For the purposes of a combinatorial perspective on Feynman graphs, the most appropriate way to set up the graphs will not have the edges or the vertices as the fundamental bits, but rather will be based on half edges. This set up is based on [1], see also [2].

Definition 12 A *graph* G is a set of half edges along with

- a set $V(G)$ of disjoint subsets of half edges known as *vertices* which partition the set of half edges, and
- a set $E(G)$ of disjoint pairs of half edges known as *internal edges*.

Those half edges which are not in any internal edge are known as *external edges*.

A *half edge labelling* of a graph with half edge set H is a bijection $H \to \{1, 2, \ldots, |H|\}$. A graph with a half edge labelling is called a *half edge labelled graph*. See Fig. 5.1 for an example.

Graphs and labelled graphs are considered up to isomorphism. Multiple edges and loops in the sense of graph theory (self loops) are allowed.

Graphs in this sense, whether labelled or not, form a combinatorial class with the size being the number of half edges. Later, in the context of combinatorial physical theories, we will put types on the half edges and restrict which types can come together at a vertex. If the degree is bounded then the number of vertices is also a notion of size under which these graphs form a combinatorial class. We will also typically consider graphs with a fixed set of external edges. Once the external edges are fixed the dimension of the cycle space of the graph (known as the *loop number* or the first Betti number; see Sect. 5.5 for more details) is another notion of size under which these graphs still form a combinatorial class. The loop number will be the most important notion of size for us.

Let's think about the interplay between labelled and unlabelled.

© The Author(s) 2017
K. Yeats, *A Combinatorial Perspective on Quantum Field Theory*,
SpringerBriefs in Mathematical Physics 15, DOI 10.1007/978-3-319-47551-6_5

Fig. 5.1 A half edge
labelled graph. The external
edges are 1, 4, 5, and 8. The
internal edges are the pairs
{2, 3}, {6, 7}, and {9, 10}

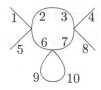

Lemma 1 *Let G be a connected graph with n half edges. Let m be the number of half edge labelled graphs giving G upon forgetting the labelling, and let Aut(G) be the automorphism group of G. Then*

$$\frac{m}{n!} = \frac{1}{|Aut(G)|}$$

Proof Aut(G) acts freely on the $n!$ half edge labellings of G. The orbits are the m isomorphism classes of half edge labellings. The result follows by elementary group theory.

The primary consequence of Lemma 1 is the following. Suppose we start with the augmented generating function of half edge labelled graphs, say $H(x)$, and the augmented generating function of the half edge graphs without labels, say $U(x)$. Next let ϕ be any evaluation map which forgets the labelling (so we can see it as defined either on the labelled or the unlabelled graphs with the same result), and finally let $\tilde{\phi}$ be the map which takes a half edge labelled graph G with n half edges to $\phi(G)/n!$ and $\widehat{\phi}$ be the map which takes a graph G to $\phi(G)/|\text{Aut}(G)|$. Then

$$\tilde{\phi}(H(x)) = \widehat{\phi}(U(x)).$$

In particular, the exponential generating function for half edge labelled graphs is identical to the ordinary generating function for half edge graphs weighted with $1/|\text{Aut}(G)|$. This factor $1/|\text{Aut}(G)|$ is called the ***symmetry factor*** of the graph.

This observation is important because it is a standard combinatorial fact that if $D(x)$ is the exponential generating function for some labelled objects and $C(x)$ is the exponential generating function for the connected objects only then

$$\exp(C(x)) = D(x)$$

while the situation is more complicated for unlabelled objects. In the perturbative expansion, Feynman graphs appear weighted by their symmetry factors. The symmetry factor is showing that Feynman graphs are in fact counted in a hidden labelled way. Furthermore, this remains true for the more complicated evaluations of the augmented generating function which explains why we get an exponential relation between amplitudes from sums over connected Feynman graphs and sums over all Feynman graphs.

Conventionally, we think of the internal structure of a Feynman graph as being unlabelled, but we weight with the symmetry factor and so get the exponential relation. The external edges, on the other hand, are usually viewed as labelled. In a drawing this is often implicitly shown by their location on the page. So for example

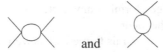 and

are viewed as different. This convention is very natural given that in a stereotypical high energy physics application the external edges are the particles from your accelerator beam and those measured by your detectors, so they are all distinguished.

Most of the time the half edges don't really matter in which case we'll think of graphs in terms of edges and vertices as usual. However, when things get unclear, the way to sort it out is to think of the half edges. For example, for some graphs it is clear how many automorphisms they have, but for others it may not be clear unless one thinks explicitly about half edges, for example

has a nontrivial automorphism involving swapping the two half edges of the self loop. As a second example it may not be clear what should count as a subgraph unless one thinks explicitly about half edges, for example

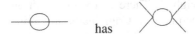 has

as a subgraph in three different ways by keeping any two of the internal edges and breaking the third into its two half edges which are then external. See Sect. 5.3 for more details on the kinds of subgraphs which matter for us.

5.2 Combinatorial Physical Theories

To make the combinatorics of renormalization Hopf algebras work, we only need a small amount of physics information, which we capture in the notion of a combinatorial physical theory.

Definition 13 A *combinatorial physical theory* is a set of half edge types along with

1. a set of pairs of not necessarily distinct half edge types defining the permissible edge types,
2. a set of multisets of half edge types defining the permissible vertex types,
3. an integer associated to each edge type and each vertex type, known as a ***power counting weight***, and

4. a nonnegative integer representing the dimension of spacetime.

A graph in a given combinatorial physical theory is a graph G as defined above with a half edge type for each half edge of G such that the pair of half edges forming any internal edge of G have types which form a permissible edge type in the theory and the multiset of half edges forming any vertex of G have types which form a permissible vertex type in the theory.

Although it is a little bit heavy, this framework is well suited to Feynman graphs. For example, oriented and unoriented edge types exist on an equal footing: unoriented edge types come from edge types made of two copies of the same half edge type while oriented edge types come from two different half edge types (a front and back half of the edge type).

Here are some standard quantum field theories in this framework.

- Quantum electrodynamics (QED) has 3 half edge types, a half photon, a front half fermion, and a back half fermion. There are two edges type, the pair of two half photons, giving a photon edge, drawn ～～～, which has power counting weight 2, and the pair of a front half fermion and a back half fermion, giving a fermion edge, drawn ——►—, which is oriented and has power counting weight 1. There is one vertex consisting of one of each half edge type and with weight 0. The dimension of spacetime is 4. The Feynman graphs in Chap. 2 are both QED graphs.

- Yukawa theory also has 3 half edge types, a half meson edge, a front half fermion edge, and a back half fermion edge. Two half mesons give a meson edge, drawn ...., with weight 2 and a pair of each half fermion gives a fermion edge, drawn ——►—, with weight 1. There is a vertex with one of each half edge and the dimension of spacetime is 4. Combinatorially, this is identical to QED. The substantial physical differences between these two theories are captured in the Feynman rules of the theories.

- Quantum chromodynamics (QCD) is the theory of the interactions of quarks and gluons. As a combinatorial physical theory it has 5 half edge types, a half gluon, a front half fermion, a back half fermion, a front half ghost, and a back half ghost. There are 3 edge types and 4 vertex types with weights as described in Table 5.1. The dimension of spacetime is again 4.

- There are two scalar field theories we want to consider. In both cases there is just one half edge type and just one edge type consisting of a pair of the half edges. This edge type has weight 2. The difference is the vertex.

ϕ^4 is the scalar theory with a 4-valent vertex, that is the vertex consists of a multiset of 4 copies of the half edge. It has weight 0. We'll take the dimension of spacetime to be 4 since that is where ϕ^4 is renormalizable. See more below on renormalizability.

ϕ^3 is the scalar theory with a 3-valent vertex, that is the vertex consists of a multiset of 3 copies of the half edge. It also has weight 0. The dimension of spacetime is 6 in order to achieve renormalizability.

Table 5.1 Edge and vertex types in QCD with power counting weights

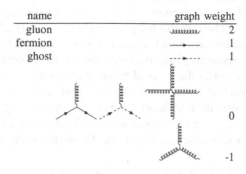

name		graph weight
gluon		2
fermion		1
ghost		1
		0
		-1

The examples of the previous section were in ϕ^4 theory.

The idea here is that the Feynman rules associate a formal integral expression to a graph by associating a factor in the integrand to each internal edge and vertex according to their type. Additionally directed cycles contribute a factor of -1 each. The information we are extracting in the weights is simply the net degree of the integration variables appearing in the factor of the integrand for that type. Along with the number of total integrations, which depends on the dimension and the loop number, we can use power counting to see how the integral behaves as all integration variables get large, that is we can understand the overall ultraviolet divergence of the integral. Specifically we can make the following definitions.

Definition 14 For a Feynman graph G in a combinatorial physical theory T, let $w(a)$ be the power counting weight of a where a is an internal edge or a vertex of G and let D be the dimension of spacetime. Then the *superficial degree of divergence* is

$$D\ell - \sum_e w(e) - \sum_v w(v)$$

where ℓ is the loop number of the graph.

If the superficial degree of divergence of a graph is nonnegative we say the graph is *divergent*. If it is 0 we say the graph is *logarithmically divergent*.

This corresponds to whether or not the integral diverges for large internal momenta and so ignoring infrared divergences it matches our earlier notion of divergence in Feynman graphs.

Given a graph in a theory, the multiset of half edge types of the external edges is the *external leg structure* of the graph. We'll be interested in the set of external leg structures which give divergent graphs.

We'll say a combinatorial physical theory T (in a given dimension) is *renormalizable* if the superficial degree of divergence of the graph depends only on the multiset of its external edges. All the example theories above are renormalizable in this sense. More typically in quantum field theory we would say a theory is renormalizable if all graphs at all loop orders can be renormalized without introducing more than finitely many new parameters. The way this happens in the theories which are important in particle physics is that there are finitely many families of divergent graphs, indexed by certain external leg structures, giving the connection with our definition. In contrast, in unrenormalizable theories there would be infinitely many families with divergent graphs and so to renormalize the whole theory would require infinitely many new parameters. Superrenormalizable theories are ones where convergence gets better as the loop order gets larger. These are not renormalizable in our sense as they are not at their critical dimension, but they are easy to deal with physically since convergence improves.

As an example, consider a scalar field theory with a k valent vertex of weight 0 and edges of weight 2. Let G be a graph in this theory with e internal edges, q external edges, and loop number ℓ. For the theory to be renormalizable we need $D\ell - 2e$ to depend only on the number of external edges. By Euler's formula along with regularity we have

$$e(2 - k) + k\ell = k - q$$

so for the theory to be renormalizable we need $D = 2k/(k - 2)$. If $k = 4$ then the theory is renormalizable in $D = 4$ while if $k = 3$ then the theory is renormalizable in $D = 6$. No other values of $k > 2$ give a theory with an integer dimension of spacetime since the only possible common factors of k and $k - 2$ are 2 and 1.

5.3   Renormalization Hopf Algebras

To build renormalization Hopf algebras we first start with all the divergent graphs in a fixed renormalizable theory. As observed in Sect. 5.1 we can restrict our attention to connected graphs since we can go from connected graphs to all graphs by exponentiating.

We say a Feynman graph is *one-particle irreducible* or *1PI* if it is connected and after removing any one internal edge it remains connected. Graph theorists call this *2-edge-connected*. Sometimes we are a little sloppy and talk about a disconnected graph being 1PI if all of its connected components are 1PI. This is what a graph theorist would call *bridgeless*. At the level of the generating function or perturbation series we can go from connected graphs to 1PI graphs by using the Legendre transform, so we can restrict to 1PI graphs.

Let $\mathscr{G}$ be the set of connected 1PI graphs in a fixed theory. As with the Connes-Kreimer Hopf algebra, renormalization Hopf algebras will be polynomial algebras first. Specifically as an algebra our Hopf algebra $\mathscr{H}$ is $K[\mathscr{G}]$. We will think of this multiplication as disjoint union and so view a monomial as a possibly disconnected graph. The counit will be the same as in the Connes-Kreimer Hopf algebra and $\mathscr{H}$ is graded by the loop order so the antipode will come for free. This means that the

only thing we need to define is the coproduct. To do so we need to write down a few
definitions about subgraphs.

For us subgraphs are full in the sense that all half edges adjacent to a vertex in
a subgraph must themselves be in the subgraph. However, if both half edges of an
internal edge are in the subgraph this does not imply that the internal edge must be
in the subgraph. For example, as noted before, we can find

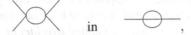 in ,

by separating the two half edges in any one of the internal edges. That is, our subgraphs are full relative to the vertices but not relative to the internal edges.

Also we will be interested in subgraphs with all their connected components 1PI.
Again note the internal edges can be cut, so for example the disconnected bridgeless
graph

is a subgraph of

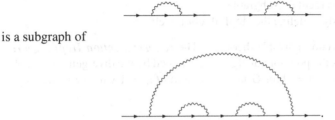

Finally we are interested in subgraphs for which each connected component is
divergent. Let $\mathscr{R}$ be the set of divergent external leg structures of the theory. If an
external leg structure has size 3 or more then it defines a vertex type which may or
may not be in the theory already. Similarly an external leg structure of size 2 defines
an internal edge type. There is a small subtlety in that the edge type should be defined
not by the two half edges themselves but by their other halves in the theory—provided
we are working only with undirected edge types made of two identical half edges
and directed edge types made of front and back halves then this amounts to the same
thing.

Augment the vertex and edge types of the theory, if necessary, so that every
divergent external leg structure is in the theory. This has already been done in all
our example theories. Physically, adding a new vertex in this way is saying that
the structure of renormalization necessitates considering some new interaction that
wasn't in the original theory. Note also that we could allow 2-valent vertex types,
but it would require some extra bookkeeping to keep track of the distinction between
2-valent vertices and internal edges. Mass renormalization requires similar extra
bookkeeping.

Now we can define contraction.

Definition 15 Let G be a Feynman graph in a theory T. Let γ be a subgraph with
each connected component 1PI and divergent. Then the **contraction** of γ, denoted
G/γ is the Feynman graph in T constructed as follows.

- Begin with G,
- for each component of γ with external leg structure of a vertex type, contract the component to a new vertex of that type, and
- for each component of γ with external leg structure of an edge type, delete the component and pair the two newly unpaired half edges into a new internal edge of this type.

Subgraphs with each connected component 1PI and divergent are known as *subdivergences*. Two subdivergences which have at least one vertex in common but for which neither is a subset of the other are called *overlapping*. For example the two copies of

inside

are overlapping as divergent ϕ^3 subgraphs.

Finally we are ready to define the Hopf algebra itself.

Definition 16 Fix a renormalizable theory T. The *renormalization Hopf algebra*, $\mathcal{H}$, associated to T is the polynomial algebra generated by the divergent 1PI graphs of T with the counit $\epsilon(G) = 0$ for G nonempty and $\epsilon(\mathbb{I}) = 1$ where $\mathbb{I}$ is the empty graph, and with coproduct

$$\Delta(G) = \sum_{\substack{\gamma \subseteq G \\ \gamma \text{ product of divergent} \\ \text{1PI subgraphs}}} \gamma \otimes G/\gamma$$

on connected Feynman graphs G and extended as an algebra homomorphism to $\mathcal{H}$. The Hopf algebra $\mathcal{H}$ is graded by the loop order.

For example in ϕ^3 theory

$$\Delta\left(-\!\!\!<\!\!>\!\!\!-\right)$$

$$= -\!\!\!<\!\!>\!\!\!- \otimes \mathbb{I} + \mathbb{I} \otimes -\!\!\!<\!\!>\!\!\!- + 2 \; \bigtriangleup \otimes -\!\!-\!\!O\!\!-\!\!- \,.$$

The antipode can be used to renormalize exactly as described in Sect. 4.3.

5.4 Insertion and the Invariant Charge

The reverse operation of contraction is insertion. If γ is a Feynman graph in a given theory with external leg structure corresponding to an edge type of the theory and G

is a Feynman graph with an internal edge e of that type, then e is called an *insertion place* for γ in G and any graph resulting from breaking edge e into its two half edges and forming new internal edges by pairing those half edges with the external edges of γ, compatibly with the theory, is an *insertion* of γ into G. If the edge type is unoriented there will be two ways to insert into a given insertion place (which may or may not be isomorphic depending on the symmetries of γ) while if the edge type is unoriented there will be only one. For example, inserting

into itself can be done in one way and gives

Similarly if γ has external leg structure corresponding to a vertex type of the theory and G is a Feynman graph with a vertex v of that type then v is called an *insertion place* for γ in G and any graph resulting from replacing the half edges of v with the external edges of γ compatibly with the theory is an *insertion* of γ into G. For example, inserting

into itself in the leftmost vertex gives

Insertion gives a pre-Lie product. Specifically let $G_1 \circ G_2$ be the sum over all insertions of G_1 into G_2; each insertion place and each possible bijection of external edges with the half edges of the insertion place should be counted. For example there are two insertion places for

into itself. If we fix the right hand insertion place then there are 4! ways to choose the bijection between the half edges of the vertex and the external half edges of the inserted graph. Of these 8 of them give

and 16 of them give

Note that some sources set insertion up in the opposite direction so the reader needs to pay attention. With our conventions, it turns out that

$$G_1 \circ (G_2 \circ G_3) - (G_1 \circ G_2) \circ G_3 = G_2 \circ (G_1 \circ G_3) - (G_2 \circ G_1) \circ G_3.$$

This is the defining property of a pre-Lie product (for more on pre-Lie products in general see [3]). As a key consequence it implies that

$$[G_1, G_2] = G_1 \circ G_2 - G_2 \circ G_1$$

is a Lie bracket. Whether one prefers to think on the Hopf side with contraction or on the Lie side with insertion is substantially a matter of taste, though the flexibility to use both is always valuable.

In analogy to the tree case we also use another notation for insertion. We would like B_+^γ to be the operator of inserting into the graph γ. As in the tree case we want to be able to use this operator to write specification equations (which will be our combinatorial Dyson-Schwinger equations) for graphs. Furthermore we would like B_+^γ to be a 1-cocycle. In simple cases it works out immediately

For example in Yukawa theory we might make the combinatorial Dyson-Schwinger equation

$$X(x) = \mathbb{I} - x B_+^{\,\overset{\frown}{}} \left(\frac{1}{X(x)} \right)$$

which is the same as the specification for plane rooted trees discussed in Sect. 3.2. This example is a key running example in [1, 2].

Unfortunately, in general things are more complicated. First, if γ is not primitive then we cannot get a 1-cocycle since

$$\Delta \left(B_+^\gamma(\mathbb{I}) \right) = \Delta(\gamma)$$

but

$$\left(\mathrm{id} \otimes B_+^\gamma \right) \Delta(\mathbb{I}) + B_+^\gamma(\mathbb{I}) \otimes \mathbb{I} = \mathbb{I} \otimes \gamma + \gamma \otimes \mathbb{I}.$$

So we restrict to γ primitive for the purposes of B_+. Overlapping divergences are still a problem. Consider

$$G = \text{\textwavy}\!\!\!\prec\!\!\!\overset{\longrightarrow}{\underset{\longrightarrow}{\langle}}\!\!\!\succ\!\!\!\text{\textwavy}$$

which can be made by inserting

$$\kappa = \quad\text{[diagram]}\quad \text{into} \quad \gamma = \quad\text{[diagram]}$$

in two different ways. Naively then we might think that $B_+^\gamma(\kappa)$ should be $2G$, but then the 1-cocycle property doesn't hold. Instead we need $B_+^\gamma(\kappa) = G$. This also means that when we use B_+^γ in building specifications we don't double count G. As discussed in the first and second sections of [4] this works out more generally and results in the following.

Definition 17 For a connected 1PI Feynman graph γ define

$$B_+^\gamma(X) = \sum_{\substack{G \text{ connected 1PI} \\ \text{Feynman graph}}} \frac{\text{bij}(\gamma, X, G)}{|X|_v} \frac{1}{\text{maxf}(G)} \frac{1}{(\gamma|G)} G$$

where

- $\text{maxf}(G)$ is the number of insertion trees corresponding to G,
- $|X|$ is the number of distinct graphs obtainable by permuting the external edges of X,
- $\text{bij}(\gamma, X, G)$ is the number of bijections of the external edges of X with an insertion place of γ such that the resulting insertion gives G, and
- $(\gamma|X)$ is the number of insertion places for X in γ.

Note that this coefficient is not a pure generating function operation since it depends on the graphs themselves not just their sizes and counts. So it makes sense at the level of augmented generating functions but not after evaluations like the ones to give ordinary or exponential generating functions.

The point is that if we sum all B_+ for primitive 1PI connected graphs with a given external leg structure, inserting into all insertion places of each, then each 1PI graph with that external leg structure occurs weighted by its symmetry factor. This property is Theorem 4 of [4] in this case.

Some of this mess can be improved by keeping everything half edge labelled for the augmented generating function—taking labelled counting seriously is useful and important. However, we still need to keep track of the number of insertion trees to deal with the overcounting so the problem of not being a pure generating function operation remains.

In gauge theories something even worse can happen: there may be overlapping subdivergences with different external leg structures. For example in QCD the graph

[diagram]

can be made by inserting

into

or inserting

into   .

This makes it impossible to use a multiplicative factor to make every B_+^γ with γ primitive into a Hochschild 1-cocycle since there will be graphs which appear in the coproduct which simply don't appear on the other side of the 1-cocycle property equation. The solution is two-fold. First of all we shouldn't work with individual primitive Feynman graphs but rather take the sum over all primitives of a given loop order. Some mismatch still remains, but for the gauge theories in high energy physics this mismatch corresponds to known quantum field theory identities between graphs: the Ward identities in QED and some of the Slavnov-Taylor identities in QCD. Hence we get our 1-cocycles after all. Thinking about it another way, we could say that we assume the 1-cocycle property and then ask what identities this requires of the graphs and find in this way a kind of combinatorial derivation of these quantum field theory identities. This phenomenon is discussed in [4] and the result is proved for QED and QCD by van Suijlekom [5]. Furthermore, the ideal of these identities is a Hopf ideal and so, whichever way we want to think about it, in the end we can mod out by this ideal and work in the resulting quotient Hopf algebra.

Another important question for us is how the number of insertion places grows with the loop number when the external leg structure is fixed. Fix a theory with only one vertex type v and with all edge types either directed edges made with distinct front and back halves or undirected edges made with two identical halves. Let d be the degree of vertex v and for each edge type e in the theory, let $n(e)$ be the total number of half edges of e (of either type in the directed case) among the half edges making up v, and let $n(v) = 1$.

Proposition 3 *Take notation as above. Let G be a 1PI connected graph in this theory with external leg structure r and loop order ℓ. Then G has $2\ell n(s)/(d-2)$ insertion places of every type $s \neq r$ and*

- *if $r = v$ then G has $2\ell n(r)/(d-2) + 1$ insertion places of type r while*
- *if $r \neq v$ then G has $2\ell n(r)/(d-2) - 1$ insertion places of type r.*

Proof This is Proposition 3.9 of [1] (see also [2]). The proof is as follows.

Let h be the number of half edges of G and let n be the number of vertices. Then $dn = h$.

Suppose first that r is an edge type. Then by Euler's formula $(h-2)/2 - n + 1 = \ell$. Substituting $dn = h$ and solving for n we get $n = 2\ell/(d-2)$. This gives the correct formula for $s = v$. For $s \neq r, v$ the number of edges of type s in G is $n \cdot n(s) = 2\ell n(s)/(d-2)$ as required, while for r there is one fewer since the external edges use up one edge-worth of half edges.

Now suppose $r = v$. By Euler's formula $(h-d)/2 - n + 1 = \ell$. Substituting $dn = h$ and solving for n we get $n = 2\ell/(d-2) + 1$. This gives the result for r while for $s \neq r$ the number of edges of type s in G is $(n-1)n(s)$ since one vertex-worth of half edges is external. $(n-1)n(s) = 2\ell n(s)/(d-2)$ as desired.

This kind of counting is nice because it means the Dyson-Schwinger equations can be put into a special form. For example in QED the system to generate all divergent 1PI graphs is

$$X^{\text{vertex}} = \mathbb{I} + \sum_{\substack{\gamma \text{ primitive with} \\ \text{vertex external} \\ \text{leg structure}}} x^{|\gamma|} B_+^{\gamma} \left(\frac{(X^{\text{vertex}})^{1+2|\gamma|}}{(X^{\text{photon}})^{|\gamma|}(X^{\text{fermion}})^{2|\gamma|}} \right),$$

$$X^{\text{photon}} = \mathbb{I} - xB_+ {\sim\!\!\bigcirc\!\!\sim} \left(\frac{(X^{\text{vertex}})^2}{(X^{\text{fermion}})^2} \right),$$

$$X^{\text{fermion}} = \mathbb{I} - xB_+ {\stackrel{\frown}{-\!\!-\!\!-}} \left(\frac{(X^{\text{vertex}})^2}{X^{\text{photon}} X^{\text{fermion}}} \right),$$

where $|\gamma|$ is the loop number of γ. Now let

$$Q = \frac{(X^{\text{vertex}})^2}{(X^{\text{photon}})^{|\gamma|}(X^{\text{fermion}})^2}.$$

Then we can rewrite the system as

$$X^{\text{vertex}} = \mathbb{I} + \sum_{\substack{\gamma \text{ primitive with} \\ \text{vertex external} \\ \text{leg structure}}} x^{|\gamma|} B_+^{\gamma}(X^{\text{vertex}} Q^{|\gamma|}),$$

$$X^{\text{photon}} = \mathbb{I} - xB_+ {\sim\!\!\bigcirc\!\!\sim} (X^{\text{photon}} Q),$$

$$X^{\text{fermion}} = \mathbb{I} - xB_+ {\stackrel{\frown}{-\!\!-\!\!-}} (X^{\text{fermion}} Q).$$

In all three equations the argument to B_+ is one extra copy of the X we are currently working with and a power of Q depending on the loop number. Q is the combinatorial avatar of the *invariant charge*.

For QCD, which has more than one vertex, the Slavnov-Taylor identities again put us back into the situation with a combinatorial invariant charge.

In the single equation case, the analogous situation is when the Dyson-Schwinger equation has the form

$$T(x) = \mathbb{I} - \sum_{k \geq 1} x^k B_+^k (T(x) Q(x)^k)$$

But now $Q(x)$ can only be some power of $T(x)$. So writing $Q(x) = T(x)^{-s}$ we get the important special case

$$T(x) = \mathbb{I} - \sum_{k \geq 1} x^k B_+^k (T(x)^{1-sk})$$

The parameter s is telling us how the number of insertion places grows as the loop order grows and will be a key parameter in the second part.

5.5 Graph Theory Tools

For the final part of this brief we will look inside primitive Feynman graphs to try to understand parts of their Feynman integrals. To do this we need some notions from graph theory. Assume G is a connected graph. A *spanning tree* of a graph G is a subset of internal edges of G which is connected, has no cycles, and is incident to every vertex of G. Using this we can define the following polynomial.

Definition 18 To each edge of G associate a variable a_e. The (dual) *Kirchhoff polynomial* or first Symanzik polynomial of a graph G is

$$\Psi_G = \sum_T \prod_{e \notin T} a_e$$

where the sum runs over all spanning trees of G.

The Kirchhoff polynomial can also be defined from a determinant. Given a graph G choose an arbitrary orientation for the edges of G and choose an order for the edges and vertices of G. Then define the *signed incidence matrix* of G to be the $V(G) \times E(G)$ matrix with i, jth entry 1 if edge j begins at vertex i, -1 if edge j ends at vertex i, and 0 otherwise. For example the cycle of length 3 has signed incidence matrix

$$\begin{bmatrix} 1 & 0 & -1 \\ -1 & 1 & 0 \\ 0 & -1 & 1 \end{bmatrix}.$$

Note that the sum of the entries of each column of the signed incidence matrix is 0. For a connected graph this is the only dependence; the rank is one less than the number of rows. Let E be any matrix obtained from a signed incidence matrix by removing one row. This matrix has the same dependence information as the original.[1] Let Λ be the diagonal matrix of the edge variables. Then

$$\det \begin{bmatrix} \Lambda & E^T \\ -E & 0 \end{bmatrix} = \Psi_G \tag{5.1}$$

This fact is essentially the matrix-tree theorem. Consider any product of variables. In the determinant this monomial will occur with coefficient the square of the determinant of the columns of E not associated to these variables. The matrix-tree theorem says that a square submatrix of E has determinant ± 1 precisely if the columns are the edges of a spanning tree and 0 otherwise. See Proposition 21 of [6] or the introduction of [7] for more details.

We will call the matrix of (5.1) the **exploded Laplacian** of the graph. The usual Laplacian of the graph is the matrix EE^T and we can augment it with variables by using $E\Lambda E^T$. Expanding $\det(E\Lambda E^T)$ by the Cauchy-Binet formula we again see the matrix-tree theorem, but this time the polynomial takes the variables in the tree not those not in the tree

$$\det(E\Lambda E^T) = \sum_T \prod_{e \in T} a_e$$

This is the polynomial usually known as the Kirchhoff polynomial, but for us the dual version defined previously is more useful.

A third way to think about the Kirchhoff polynomial is as a variant of the multivariate Tutte polynomial [8]. Anything coming from the multivariate Tutte polynomial must be definable by a contraction deletion relation. The contraction deletion relation in this case is

$$\Psi_G = a_e \Psi_{G \setminus e} + \Psi_{G/e}$$

for any non-loop non-bridge edge e, which can be justified directly from the original spanning tree definition. Here $G \setminus e$ is G with edge e **deleted** and G/e is G with edge e **contracted**.

Another important graph theory idea is the notion of the cycle space of a graph. Consider any two cycles in a graph in terms of the edges making them up. If we take the symmetric difference of these edges we get a set of edges which is itself one or more cycles in the graph. This means that the cycles of the graph span a vector space over the field with two elements; this vector space is called the **cycle space** of the graph. The dimension of the cycle space is the loop number of the graph. There are many ways to think about the loop number—it is the number of edges not in a spanning tree or equivalently the minimum number of edges of the graph that can be cut to give a graph with no cycles.

---

[1] To make this precise we'd need to move into matroids. Much of what we do with Feynman graphs works very naturally with regular matroids and even some more general matroids, but that's another story.

Fig. 5.2 The *solid* edges give a planar embedding of a graph G; the *dotted* edges give the dual graph G^*

Finally, we need the notion of planar duality. A graph is *planar* if it can be drawn on the plane with each vertex a distinct point and each edge a curve where the edges only intersect at their ends. Such a drawing of a planar graph is a *planar embedding*. Cutting on the edges divides the plane into regions which are the *faces* of the embedding, including the one *external* or infinite face. Given a planar embedding of a graph we can define the *planar dual* as the graph with a vertex for each face of the embedding (including the external face) and where each edge of the original graph contributes an edge of the dual joining the vertices corresponding to the faces on either side of the original edge. Thus bridges in the original graph become loops in the dual. We will use the notation G^* for the dual of G. See Fig. 5.2 for an example. To obtain an analogous dual for nonplanar graphs one must move outside the world of graphs and into matroids, but only into a very tame class of matroids, regular matroids.

5.6 Feynman Rules

Finally we need to get from graphs to physics.

In a typical quantum field theory presentation a physical theory might be defined by its *Lagrangian* $\mathscr{L}$. For example, for ϕ^4

$$\mathscr{L} = \frac{1}{2}\partial^\mu\phi\partial_\mu\phi - \frac{1}{2}m^2\phi^2 - \frac{\lambda}{4!}\phi^4.$$

There is one term for each vertex and edge of the theory and an additional term for each massive particle. In this case $\frac{1}{2}\partial^\mu\phi\partial_\mu\phi$ is the term for the edge of ϕ^4, $-\frac{1}{2}m^2\phi^2$ is the mass term, and $-\frac{\lambda}{4!}\phi^4$ is the term for the vertex. One can read a lot of physics directly off of the Lagrangian.

The Feynman rules can be derived from the Lagrangian in a variety of ways to suit different tastes, for instance directly [9, p. 16]; or by expanding the path integral in the coupling constant, see for example [10]. The idea of expanding the path integral is as follows. If we integrate the Lagrangian over spacetime we get the *action*, schematically $S = \int d^D y \mathscr{L}$ where D is the dimension of spacetime. This is in direct analogy to the situation in classical mechanics and in quantum mechanics. What about the fields? In our example there is just the one field ϕ. The (Euclidean) path integral is the integral of e^{-S} over all possibilities for the fields, schematically $\int d\phi e^{-S}$. To get things going we also want to include a source term which is linear in ϕ, say $J\phi$, so we get

$$Z = \int d\phi e^{-\int d^D y \mathscr{L} + J\phi}$$

Now being good physicists or good combinatorialists (but not good analysts since this isn't well defined as an integral) we expand in the coupling constant (λ in this case) and in J and see what happens. If we first expand in the coupling constant we will get, in our example, terms of the form

$$\lambda^n \int d\phi e^{-J\phi}\text{-terms quadratic in } \phi \; \phi^{4n}$$

which can be rewritten formally as J derivatives of Gaussian integrals

$$\lambda^n \frac{d^{4n}}{dJ^{4n}} \int d\phi e^{-J\phi}\text{-terms quadratic in } \phi$$

Next expand in J. The derivatives came from the extraneous powers of ϕ which came from four times the power of the coupling constant in a given term in the coupling constant expansion. Each derivative needs a power of J to consume in order to get a nonzero answer. This means that each power of the coupling constant needs to be matched up with four Js, or said another way each vertex needs to be matched up with four half edges. The quadratic terms in ϕ similarly match pairs of half edges into internal edges. This matching up is called Wick contraction and what it is doing is building Feynman graphs out of half edges. Any extra powers of J are the unmatched half edges, that is external edges. The remaining bit of integral carried along with any particular graph is the Feynman integral associated to this graph. See for instance [10]. If spacetime is zero dimensional this idea is rigorous and is used by both combinatorialists and physicists for counting graphs, see for example [11, 12].

Gauge theories are a bit more complicated. Geometrically rather than being defined directly on spacetime they are defined on a fibre bundle over spacetime. The structure group of the fibre bundle is called the gauge group. A gauge field (for example the photon in QED or the gluon in QCD) is a connection. A *gauge* is a local section. See for example [13, Chap. 15]. If we don't want to work geometrically then we need to choose a gauge.

There are many ways to choose a gauge each with different advantages and disadvantages. For the present purpose we're interested in a 1-parameter family of Lorentz covariant gauges called the R_ξ gauges. The parameter for the family is denoted ξ and is the ξ. The R_ξ gauges can be put into the Lagrangian in the sense that in these gauges we can write a Lagrangian for the theory which depends on ξ. For example, for QED in the R_ξ gauges we have (see for example [14, p. 504])

$$\mathscr{L} = -\frac{1}{4}(\partial_\mu A_\nu - \partial_\nu A_\mu)^2 - \frac{1}{2\xi}(\partial_\mu A^\mu)^2 + \bar{\psi}(i\gamma^\mu(\partial_\mu - ieA_\mu) - m)\psi$$

where the γ^μ are the Dirac gamma matrices (see [13, Sect. 3.2]). Once the gauge is in the Lagrangian we can follow the same Feynman diagram or path integral story as above.

Whether we want to take the Feynman rules as basic input information or as derived, the next step is to think about what Feynman rules do combinatorially and physically.

I will choose not to include the coupling constant with the Feynman rules. This is because in our augmented generating functions we already have a counting variable x which counts the loop number of the graph and ultimately is the coupling constant. In theories with more than one coupling constant we could either keep a preferred one in the augmented generating functions or use multivariate augmented generating functions.

Feynman rules associate to each Feynman graph (in a physical theory) a formal integral, that is, an integrand and a space to integrate over but with no assurances that the resulting integral is convergent. We'll work explicitly with Euclidean momentum space integrals in this part and the second part of this brief. In the final part we will work more with parametric integrals. See Chap. 11 for more on the interplay between different representations.

Momentum space Feynman rules are structured as follows. They are rules which tell us how each edge and vertex gives a piece of the integrand of the formal integral. The basic shape of these formal integrals is

$$\int_{\mathbb{R}^{D|v_\gamma|}} \mathrm{Int}_\gamma \prod_{k \in v_\gamma} d^D k$$

where D is the dimension of spacetime and v_γ is a finite index set corresponding to the set of integration variables appearing in Int_γ.

To build Int_γ, associate to each half edge of γ a tensor index, associate to each internal and external edge of γ a variable (the momentum, with values in $\mathbb{R}^D$) and an orientation of the edge with the restriction that for each vertex v the sum of the momenta of edges entering v equals the sum of the momenta of edges exiting v. If we view the variables associated to the external edges as fixed then the $\mathbb{R}$-vector space of the remaining free edge variables has dimension the loop number of the graph. Let v_γ be a basis of this vector space. Let Int_γ be the product of the Feynman rules applied to the type of each external edge, internal edge, and vertex of γ, along with the assigned tensor indices, the edge variables as the momenta, and a factor of -1 for each fermion cycle. Int_γ depends on the momenta $q_1, \ldots, q_n$ for the external edges; these variables are not "integrated out" in the formal integral.

The piece associated to an edge or vertex of a graph we will call a **tensor expression** meaning an expression for a tensor written in terms of the standard basis for $\mathbb{R}^D$ where D is the dimension of spacetime. Such expressions will be intended to be multiplied and then interpreted with Einstein summation. An example of a tensor expression in indices μ and ν is

$$\frac{g_{\mu,\nu} - \xi \frac{k_\mu k_\nu}{k^2}}{k^2}$$

where g is the Euclidean metric, $k \in \mathbb{R}^4$, k^2 is the standard dot product of k with itself, and ξ is the gauge. Such a tensor expression is meant to be a factor of a larger expression like

$$\gamma_\mu \frac{1}{\not k + \not p - m} \gamma_\nu \left(\frac{g_{\mu,\nu} - \xi \frac{k_\mu k_\nu}{k^2}}{k^2} \right) \tag{5.2}$$

where the γ_μ are the Dirac gamma matrices, $\not k$ is the Feynman slash notation, namely $\not k = \gamma^\mu k_\mu$, and m is a variable for the mass. In this Example (5.2) is the integrand for the Feynman integral for the QED graph

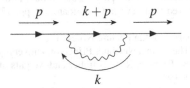

The simplest case is when there are no tensor indices. This is the case of scalar field theories. For example consider ϕ^4 with Euclidean Feynman rules, see [15, p. 268]. The Feynman rules in this case say that an edge labelled with momentum k is associated to the factor $1/(k^2 + m^2)$, where the square of a vector means the usual dot product with itself and m is the mass of the particle. The Feynman rules say that the vertex is associated to -1 (if the coupling constant λ was included in the Feynman rules the vertex would be associated with $-\lambda$). Consider

oriented from left to right with the momenta associated to the two right hand external edges summing to p and hence the momenta associated to the two left hand external edges also summing to p. Then the integral associated to this graph is

$$\int d^4 k \frac{1}{(k^2 + m^2)((p + k)^2 + m^2)}$$

where $d^4 k = dk_0 dk_1 dk_2 dk_3$ with $k = (k_0, k_1, k_2, k_3)$ and squares stand for the standard dot product.

More on Feynman rules and building Feynman integrals can be found in any quantum field theory textbook, for example [10, 13, 15].

References

1. Yeats, K.: Rearranging Dyson-Schwinger equations. Mem. Am. Math. Soc. **211** (2011)
2. Yeats, K.A.: Growth estimates for Dyson-Schwinger equations. Ph.D. thesis, Boston University (2008)
3. Chapoton, F., Livernet, M.: Pre-Lie algebras and the rooted trees operad. Int. Math. Res. Not. **8**(8), 395–408 (2001). arXiv:math/0002069
4. Kreimer, D.: Anatomy of a gauge theory. Ann. Phys. **321**, 2757–2781 (2006). arXiv:hep-th/0509135v3
5. van Suijlekom, W.D.: Renormalization of gauge fields: a Hopf algebra approach. Commun. Math. Phys. **276**, 773–798 (2007). arXiv:hep-th/0610137
6. Brown, F.: On the periods of some Feynman integrals. arXiv:0910.0114
7. Vlasev, A., Yeats, K.: A four-vertex, quadratic, spanning forest polynomial identity. Electron. J. Linear Alg. **23**, 923–941 (2012). arXiv:1106.2869
8. Sokal, A.D.: The multivariate Tutte polynomial (alias Potts model) for graphs and matroids. In: Webb, B.S. (ed.) Surveys in Combinatorics, pp. 173–226. Cambridge (2005). arXiv:math/0503607
9. Cvitanović, P.: Field Theory. Nordita Lecture Notes (1983)
10. Zee, A.: Quantum Field Theory in a Nutshell. Princeton University Press, Princeton (2003)
11. Cvitanović, P., Lautrup, B., Pearson, R.B.: Number and weights of Feynman diagrams. Phys. Rev. D **18**(6), 1939–1949 (1978)
12. Lando, S.K., Zvonkin, A.K.: Graphs on Surfaces and Their Applications. Springer, Berlin (2004)
13. Peskin, M.E., Schroeder, D.V.: An Introduction to Quantum Field Theory. Westview, Boulder (1995)
14. Cheng, T.P., Li, L.F.: Gauge Theory of Elementary Particle Physics. Oxford University Press, Oxford (1984)
15. Itzykson, C., Zuber, J.B.: Quantum Field Theory. McGraw-Hill (1980). Dover edition (2005)

Part II
Dyson-Schwinger Equations

Chapter 6
Introduction to Dyson-Schwinger Equations

From our graphs-first approach, Dyson-Schwinger equations begin as functional equations for the augmented generating functions of Feynman graphs. At this level these equations are purely combinatorial objects, the ***combinatorial Dyson-Schwinger equations*** of the previous sections. More strictly, and to match a more standard physics treatment, combinatorial Dyson-Schwinger equations should be functional equations for the augmented generating functions of families of 1PI Feynman graphs with a fixed external leg structure, or their equivalents at the level of insertion trees.

Applying Feynman rules to the combinatorial Dyson-Schwinger equations we get functional equations for the augmented generating functions evaluated with the Feynman rules. That is, we get functional equations for the ***Green functions***. If we used arbitrary combinatorial constructions in our original generating function functional equations, then there would be no guarantee that the equations after applying Feynman rules would have any reasonable analytic expression. Fortunately, keeping to the stricter setting of physically reasonable combinatorial Dyson-Schwinger equation, we can build the combinatorial Dyson-Schwinger equations using B_+ operators and composition with other formal power series (mostly polynomials and Seq—so polynomials and geometric series). These operations play well with Feynman rules. Composition with a formal power series remains unchanged since the Feynman rules are algebra homomorphisms (see Sect. 4.4). What the Feynman rules do with a B_+ is, by the universal property of Sect. 4.4, replace it by another operator, typically a formal integral operator. The toy Feynman rules of Chap. 4 modelled this.

The consequence is that applying Feynman rules to the combinatorial Dyson-Schwinger equations gives what we will call ***analytic Dyson-Schwinger equations***, for now still viewed as formal objects rather than honestly analytic objects. These are formal objects with an analytic flavour as they are integral equations for the Green functions. Next we want to ask what we can learn about the Green functions from the analytic Dyson-Schwinger equations. Staying combinatorial, this typically

© The Author(s) 2017
K. Yeats, *A Combinatorial Perspective on Quantum Field Theory*,
SpringerBriefs in Mathematical Physics 15, DOI 10.1007/978-3-319-47551-6_6

means asking about series solutions to these integral equations which have nice interpretations themselves as sums over discrete objects.

The final step from this viewpoint is to see to what extent the formalness of the objects can be dropped. We won't discuss this much in this brief as it departs from the combinatorial side of things. One powerful but difficult way to approach this is still to start with series solutions (whether indexed by nice discrete objects or not) and then look at resummation techniques, for example see [1] and the references therein. We can also look for special cases where the Dyson-Schwinger equations may be amenable to analysis as differential equations, for example see [2, 3]. Notably in [2] this viewpoint shows a potential way to avoid a Landau pole in QED.

A nice example of how this all goes together comes from all ways of inserting

into itself in massless Yukawa theory. This was studied by Broadhust and Kreimer in [4] and is a running example in [5, 6].

Combinatorially this corresponds to the equation

$$X(x) = \mathbb{I} - x B_+ \underset{}{\overset{}{\quad}} \left(\frac{1}{X(x)} \right).$$

Applying the Feynman rules of Yukawa theory, see [7], this becomes

$$G(x, L) = 1 - \frac{x}{q^2} \int d^4k \, \frac{k \cdot q}{k^2 G(x, \log k^2)(k + q)^2} - \cdots \Big|_{q^2 = \mu^2}$$

where the $\cdots |_{q^2 = \mu^2}$ means take the same integrand but evaluate it at $q^2 = \mu^2$, that is we are renormalizing by subtraction. This is more or less a recognizable Dyson-Schwinger equation, having only nonstandard notation and normalization. Here x is playing the role of the coupling constant, q is the momentum going through, renormalization is taking place by subtraction at a fixed reference μ, $G(x, L)$ is the fermion Green function, and the tree level term has been normalized to 1. The Feynman rules took the augmented generating function $X(x)$ to the Green function $G(x, L)$ and they took the B_+ to the integral operator $\frac{1}{q^2} \int d^4k \frac{k \cdot q}{k^2 (k+q)^2}$ which is the loop integral for the outermost copy of

Note that the recursive appearance of $X(x)$ on the right became the $G(x, \log k^2)$ on the right.

Broadhust and Kreimer solve this equation, after substantial work, by recognizing the series expansion as being built from the asymptotic expansion of the complementary error function.

One outstanding question in the reader's mind may be how this connects to the usual presentation of Dyson-Schwinger equations in quantum field theory. The classical Euler-Lagrange equations for the field can be upgraded to equations which hold for all the Green functions of the field, see [7, Sect. 9.6]. These are the Dyson-Schwinger equations and this explains why they are the quantum equations of motion, namely because they are analogous to the classical Euler-Lagrange equations.

Obtained this way, the equations hold nonperturbatively and so in principle their solutions give the Green functions nonperturbatively. Alternately they can be expanded to give the Dyson-Schwinger equations at the level of Feynman diagrams—these are the analytic Dyson-Schwinger equations we have been working with above.

We will take a layered approach to studying Dyson-Schwinger equations. First we will think about them purely combinatorially. Then we will add in simple Feynman rules like the tree factorial Feynman rules. Next we will pass to more realistic Feynman rules but still think of the solutions to Dyson-Schwinger equations strictly in terms of their series expansions. Finally we will look at some new approaches to the log expansions using different underlying combinatorial objects.

References

1. Klaczynsky, L.: Resurgent transseries & Dyson-Schwinger equations. arXiv:1601.04140
2. van Baalen, G., Kreimer, D., Uminsky, D., Yeats, K.: The QED beta-function from global solutions to Dyson-Schwinger equations. Ann. Phys. **234**(1), 205–219 (2008). arXiv:0805.0826
3. van Baalen, G., Kreimer, D., Uminsky, D., Yeats, K.: The QCD beta-function from global solutions to Dyson-Schwinger equations. Ann. Phys. **325**(2), 300–324 (2010). arXiv:0805.0826
4. Broadhurst, D., Kreimer, D.: Exact solutions of Dyson-Schwinger equations for iterated one-loop integrals and propagator-coupling duality. Nucl. Phys. B **600**, 403–422 (2001). arXiv:hep-th/0012146
5. Yeats, K.: Rearranging Dyson-Schwinger equations. Mem. Am. Math. Soc. **211** (2011)
6. Yeats, K.A.: Growth estimates for Dyson-Schwinger equations. Ph.D. thesis, Boston University (2008)
7. Peskin, M.E., Schroeder, D.V.: An introduction to quantum field theory. Westview (1995)

Chapter 7
Sub-Hopf Algebras from Dyson-Schwinger Equations

7.1 Simple Tree Classes Which Are Sub-Hopf

The defining equation for any simple tree class is a combinatorial Dyson-Schwinger equation, see Sect. 3.2, so all of the combinatorics of simple trees could be viewed as part of the combinatorics of Dyson-Schwinger equations. We would like to be a bit more focused and consider what sort of questions regarding simple tree classes come naturally from the physics and what sort of mathematical insights about simple tree classes give physically interesting results.

One important family of results which speaks to both points has been proved by Loïc Foissy in [1–3], see also the lecture notes [4]. This chapter is a summary of some of Foissy's results. Thinking specifically about Dyson-Schwinger equations, take the solution to a combinatorial Dyson-Schwinger equation, which is some augmented generating function, and then we can ask when the polynomial algebra generated by the coefficients of this augmented generating function is in fact a Hopf algebra. Foissy investigated this question in different cases. Once we are equipped with an answer to this question we can ask what the physical meaning of these special Hopf Dyson-Schwinger equations is. The simplest case which illustrates many of the general methods is the case of single Dyson-Schwinger equations with a single B_+ in the Connes-Kreimer Hopf algebra $\mathcal{H}$, that is the case of simple tree classes. Foissy characterized the sub Hopf algebras in this case in [1] and the answer is interesting and quite pretty.

The setup is as follows, see [1] or [4] for more details. Suppose we have a combinatorial Dyson-Schwinger equation of the form

$$T(x) = x B_+(f(T(x))) \tag{7.1}$$

where $f(z) \in K[[z]]$ with $f(0) = 1$. First observe that there is a unique solution

$$T(x) = \sum_{n \geq 1} t_n x^n$$

© The Author(s) 2017
K. Yeats, *A Combinatorial Perspective on Quantum Field Theory*,
SpringerBriefs in Mathematical Physics 15, DOI 10.1007/978-3-319-47551-6_7

with $t_n \in \mathcal{H}$ to this combinatorial Dyson-Schwinger equation. Uniqueness is simply a restatement of how tree specifications work—larger trees are recursively built up out of the smaller ones and t_n is the sum of all of them of size n. Note that the t_n will have trees with multiplicities. This corresponds to the fact that if we are thinking of this as a simple tree class then f is determining some combinatorial structure on the children, but when we forget that structure then we can end up with multiple copies of the tree in $\mathcal{H}$. Writing down what this recursive construction is directly, if $f(z) = 1 + \sum_{n \geq 1} a_n z^n$ then we get (see Proposition 18 of [4])

$$t_0 = 0$$
$$t_1 = \bullet$$

$$t_{n+1} = \sum_{k=1}^{n} \sum_{i_1 + \cdots + i_k = n} a_k B_+(t_{i_1} \ldots t_{i_k})$$

Now consider the polynomial algebra $K[t_1, t_2, t_3, \ldots]$. This is certainly a subalgebra of $\mathcal{H}$. The question is when $K[t_1, t_2, t_3, \ldots]$ is sub-Hopf. As is typical, the counit works trivially, so the question really is whether or not

$$\Delta(K[t_1, t_2, t_3, \ldots]) \subseteq K[t_1, t_2, t_3, \ldots] \otimes K[t_1, t_2, t_3, \ldots]$$

If this is true we say f or the combinatorial Dyson-Schwinger equation (7.1) is Hopf.

This is not just some esoteric algebraist's question. The series $T(x)$ is standing in (via insertion trees) for the sum over all Feynman diagrams contributing to some process of interest. If $K[t_1, t_2, t_3, \ldots]$ is closed under the coproduct then that tells us that it makes sense mathematically to apply all the renormalization Hopf algebra machinery directly on $K[t_1, t_2, t_3, \ldots]$ and so we can renormalize at the level of the Green function rather than Feynman diagram by Feynman diagram. Now this isn't yet quite right physically since usually some finite number of terms need renormalizing, not just one. This corresponds to the fact that to generate all the divergent diagrams you would need not just one Dyson-Schwinger equation but a system of them. These more physical situations are the topic of the next section.

As it turns out there is precisely a 2-parameter family of Hopf combinatorial Dyson-Schwinger equations.

Theorem 2 (Theorem 24 of [4]) *Let* $f(z) = 1 + \sum_{n \geq 1} a_n z^n$. *The following are equivalent:*

1. *The combinatorial Dyson-Schwinger equation (7.1) is Hopf.*
2. *$f(z)$ is one of the following*

$$\begin{cases} f(z) = 1 \\ f(z) = e^{\alpha z} & \alpha \neq 0 \\ f(z) = (1 - \alpha\beta z)^{-\frac{1}{\beta}} & \alpha\beta \neq 0 \end{cases}.$$

In fact slightly more than this holds. The analogous statement can be made for plane rooted trees and noncommutative polynomial algebras but this extra generality only gives the same possible solutions (see [1]). Foissy gives two proofs of this result. The first, given in [1], plays recursively with the trees in order to obtain a differential equation for f which is then solved to obtain the possible forms of f. The other, given in [4], is more algebraic and uses the pre-Lie insertion structure defined in Sect. 5.4. In both cases the structure coefficients $n(t, t')$ are a key tool. Let t and t' be trees then

$$n(t, t') = \text{number of leaves of } t \text{ which give } t' \text{ when removed.}$$

Consider the physical implications of Theorem 2. With $\beta > 0$ we are getting some power of Seq. This is the shape of Dyson-Schwinger equation we would expect when there are propagator insertions since we can insert any sequence of propagator corrections into a propagator. The power is telling us how many insertion places there are. With $\beta < 0$ we are essentially getting powers of $1 + T(x)$. This is what we expect when there are vertex insertions since we can insert either nothing (1) or any single vertex correction ($T(x)$) into a vertex. What about $f(z) = e^{\alpha z}$? Combinatorially this is telling us to take a labelled set of trees of $T(x)$ to form the children of a new tree. At the level of Feynman graphs this would say to insert any labelled set of graphs into a primitive graph. As we saw in Sect. 5.1 we can ignore the labelling in favour of weighting by symmetry factors as is standard in quantum field theory. So this again seems very natural physically.

7.2 More Physical Situations

There are two main ways in which the simple tree case is simpler than what we see in quantum field theory. First of all, simple tree classes are defined by a single equation rather than a system of equations. Second, simple tree classes have a single B_+ which adds one to the size of the tree whereas in quantum field theory we may insert into primitives at higher loop orders as well. At the level of insertion trees this corresponds to rooted trees where the vertices are weighted (corresponding to the loop number of the graph being inserted into) with positive integer weights and for each $j > 0$ there is a B_+ which adds a new root of size j. We could also ask for more than one coupling constant, see [5].

Systems of combinatorial Dyson-Schwinger equations where all the B_+ have weight 1 were studied by Foissy in [2], see also [4]. The set up is as follows. Consider rooted trees which are decorated by $\{1, 2, \ldots, n\}$ in the sense that a tree comes with a map from its vertices to $\{1, 2, \ldots, n\}$ but where the size of a tree is just the number of vertices, so each of these decorations is of size 1. We have the same Hopf algebra structure for decorated rooted trees by simply keeping track of the decorations in the constructions we used before. The systems of combinatorial Dyson-Schwinger equations studied in [2] are those of the form

$$T_i(x) = x B_+^i (f_i(T_1(x), T_2(x), \ldots, T_n(x)))$$

where each $f_i \in K[[z_1, z_2, \ldots, z_n]]$ is nonconstant. Once again the system has a unique solution determined recursively. Write $T_i(x) = \sum_{j \geq 1} t_{i,j} x^j$. Then the question is, as before, whether the polynomial algebra $K[t_{i,j}, 1 \leq i \leq n, 1 \leq j]$ is a sub Hopf algebra of the Hopf algebra of decorated rooted trees.

The solution to this problem is much more intricate than in the single equation case. The first tool we need is a restricted dependency digraph of the system. To build this write

$$f_i(z_1, z_2, \ldots, z_n) = \sum a_{j_1, j_2, \ldots, j_n}^{(i)} z_1^{j_1} \cdots z_n^{j_n}$$

and let $a_j^{(i)}$ be shorthand for $a_{0, \ldots, 0, 1, 0, \ldots, 0}^{(i)}$ with the 1 in the subscript in the jth position. Now build the directed graph with a vertex for each equation of the system and a directed edge from vertex i to vertex j if $a_j^{(i)} \neq 0$. In this case we'll write $i \to j$. This digraph is like the usual dependency digraph for a system where there is a directed edge between i and j if equation i depends on equation j, but the dependence must be linear. Because of the restrictions of being a sub Hopf algebra many relations between the coefficients are forced and so it works well to use this restricted dependency digraph instead of the usual dependency digraph.

Foissy gets his characterization of sub Hopf systems by finding two special kinds of systems to use as building blocks and then using four operations to glue them together, see [4] for examples and the precise statement. The operations are

• scaling the functions f_i or their arguments (called change of variables in the nonzero case, and restriction when setting some to 0),
• concatenating independent systems,
• extending a system with a new f_0 with the form $f_0 = 1 + \sum_{i=1}^n a_i^{(0)} z_i$, and
• a kind of linear refinement called dilatation.

Dilatation is defined as follows. Begin with a system associated to the list of formal series $(f_i)_{i \in I}$. Then a dilatation of this system is a system associate to a list of formal series $(g_j)_{j \in J}$ where J can be partitioned as $J = \bigcup_{i \in I} J_i$ and for all $i \in I$ and $j \in J_i$ we have

$$g_j = f_i \left(\sum_{s \in J_1} z_s, \sum_{s \in J_2} z_s, \ldots, \sum_{s \in J_n} z_s \right)$$

The first of Foissy's building blocks is fairly simple. Start with a system with dependency digraph an oriented cycle then up to change of variables either $f_i = 1 + z_{j : i \to j}$ or $f_i = 1/(1 - z_{j : i \to j})$. Any system built from an oriented cycle with the $f_i = 1 + z_{j : i \to j}$ by applying the operations of dilatation and change of variables is called quasi cyclic. This is the first type of building block.

The second building block is more complicated. It is a system where the equations each depend on the others by a product of factors where each factor is a function of only one z_i and the function is one of the functions which was allowed in the simple tree case. There are some additional restrictions on the shape of the dependency digraph. See [4] Definition 51 for details.

We see that the situation for systems is considerably more complicated than in simple trees, but the parts relate via the same physically reasonable forms (powers of $1 + T(x)$, geometric series, and exp).

Now what if we allow B_+ with different degrees? Foissy studied this case in [3]. For a single equation this means we are looking at Dyson-Schwinger equations of the form

$$T(x) = \sum_{j \in J} x^j B_+^j(f_j(T(x)))$$

for some finite set of positive integers J where now B_+^j adds a new root of size j and each $f_j(0) = 1$.

The answer ends up being that the Dyson-Schwinger equation is Hopf in two cases. The first is the obvious generalization of the simple tree case; there exist constants λ and μ such that

$$f_j(z) = (1 - \mu z)Q(z)^j$$

where

$$Q(z) = \begin{cases} (1 - \mu z)^{-\lambda/\mu} & \text{if } \mu \neq 0 \\ e^{\lambda z} & \text{if } \mu = 0 \end{cases}.$$

Note that Q is playing the role of the invariant charge as in Sect. 5.4.

The second case is based on a divisibility condition; there exists a nonnegative integer m and a nonzero constant α such that

$$f_j(z) = \begin{cases} 1 + \alpha z & \text{if } m \mid j \\ 1 & \text{otherwise} \end{cases}.$$

This case doesn't show up in quantum field theory so far as I am aware.

Systems with B_+ of different degrees end up being a mutual extension with a similar flavour. For details see [3].

References

1. Foissy, L.: Faà di Bruno subalgebras of the Hopf algebra of planar trees from combinatorial Dyson-Schwinger equations. Adv. Math. **218**(1), 136–162 (2007). arXiv:0707.1204
2. Foissy, L.: Classification of systems of Dyson-Schwinger equations in the Hopf algebra of decorated rooted trees. Adv. Math. **224**(5), 2094–2150 (2010). arXiv:0909.0358
3. Foissy, L.: General Dyson-Schwinger equations and systems. Commun. Math. Phys. **327**(1), 151–179 (2014). arXiv:1112.2606

4. Foissy, L.: Pre-Lie algebras and systems of Dyson-Schwinger equations. In: Dyson-Schwinger Equations and Fa di Bruno Hopf Algebras in Physics and Combinatorics – DSFdB2011. IRMA Strasbourg, European Mathematical Society (2016)
5. Foissy, L.: Mulitgraded Dyson-Schwinger systems. arXiv:1511.06859

Chapter 8
Tree Factorial and Leading Log Toys

To move closer to the physical situation we need to apply Feynman rules to our combinatorial Dyson-Schwinger equations. For a first step in this direction let's consider the tree factorial Feynman rules, defined in Sect. 3.1, and similar toy Feynman rules. Specifically, we will look at tree classes and at Feynman rules which can be written in the form

$$\phi(t) = \prod_{v \in V(t)} B_{|t_v|}$$

where t_v is the subtree rooted at v, the size of a tree is its number of vertices, and $B = (B_1, B_2, B_3, \ldots)$ is a sequence called the **hook weight sequence**. As usual ϕ is extended as an algebra homomorphism to the Connes-Kreimer Hopf algebra. Feynman rules built in this form will be called **hook weight Feynman rules** and $\phi(t)$ will be called the **hook weight** of t.

To put it into the shape of the universal property form of Feynman rules (see Sect. 4.4), we can write hook weight Feynman rules as

$$\phi(B_+(f)) = B_{|f|+1}\phi(f)$$

for any forest f.

If $B_i = \frac{1}{i}$ then we have $\phi(t) = \frac{1}{t!}$. These are the tree factorial Feynman rules, except that we have left out the variable z from (3.1) since z there counts the size of the tree and the x in the combinatorial Dyson-Schwinger equation does the same, so having both is unnecessary. Another way to think about this is that hook weight Feynman rules only capture the case when the power of the coupling constant is equal to the power of the scale variable coming from the Feynman rules, that is they are giving the leading log part of the full Dyson-Schwinger equations. For more on log expansions see Chap. 10.

Note that in the universal property result, Theorem 1 of Sect. 4.4, there were stronger conclusions if the map on the level of the target algebra was itself a 1-

© The Author(s) 2017
K. Yeats, *A Combinatorial Perspective on Quantum Field Theory*,
SpringerBriefs in Mathematical Physics 15, DOI 10.1007/978-3-319-47551-6_8

cocycle. In this case the target algebra is $K[x]$ and given a hook weight series B the corresponding map on $K[x]$, called L_B, is defined by

$$L_B(x^n) = B_{n+1}x^{n+1}.$$

The natural coalgebra structure on $K[x]$ is given by taking $\Delta(x) = x \otimes 1 + 1 \otimes x$ and extending as an algebra homomorphism. Then we get the following proposition

Proposition 4 [1, Proposition 5.4] *Let B be a hook weight sequence. If*

$$L_B \Delta = (id \otimes L_B)\Delta + L_B \otimes 1$$

then $B_n = \frac{c}{n}$ for some $c \in \mathbb{K}$ and hence gives a multiple of the tree factorial Feynman rules.

Proof Simply calculate and compare coefficients

$$B_{n+1} \sum_{i=0}^{n+1} \binom{n+1}{i} x^i \otimes x^{n+1-i} = L_B \Delta(x^n) = (id \otimes L_B)\Delta + L_B \otimes 1$$

$$= \sum_{i=0}^{n} \binom{n}{i} B_{n-i+1} x^i \otimes x^{n-i+1} + B_{n+1}x^{n+1} \otimes 1$$

So

$$B_{n+1} \binom{n+1}{i} = B_{n-i} \binom{n}{i}$$

for all $0 \leq i \leq n$, and so in particular with $i = n$, $B_{n+1}(n+1) = B_1$ giving the result.

This is another aspect of the specialness of the tree factorial.

None-the-less allowing ourselves to work with all hook weight Feynman rules we do get some nice results. Perhaps the most interesting thing is that these objects were independently studied in pure combinatorics, see for example [2–4], and from a B-series perspective [5].

An early example on the pure combinatorics side of the kind of result one has is the identity due to Postnikov [2]

$$\sum_{t \text{ binary tree}|t|=n} n! \prod_{v \in V(t)} \left(1 + \frac{1}{|t_v|} \right) = 2^n(n+1)^{n-1},$$

where the binary trees are those generated by the combinatorial Dyson-Schwinger equation $T(x) = \mathbb{I} + x B_+(T(x)^2)$. Each tree in the sum is weighted by its hook weight using the hook weight sequence $B_n = 1 + 1/n$. The left hand side counts bicoloured labelled forests. If we multiplied by x^n and summed each side for $n \geq 0$ then on the left we would have a generating function for binary rooted trees weighted by these hook weights and on the right we would have the generating function for bicoloured labelled forests.

This illustrates the kind of result people are interested in. We have a class of rooted trees (in this case binary trees) and a hook weight sequence and we want to know when the generating function for the class of trees weighted by the hook weights is itself a nice sequence where nice can either mean that it is also a generating function for some reasonable combinatorial objects or can mean that it has a closed form. In quantum field theory language this says we have a combinatorial Dyson-Schwinger equation in the Connes-Kreimer Hopf algebra and some hook weight Feynman rules and we want to know when the Green function is nice either in the sense of having a direct combinatorial interpretation or having a closed form.

Algebraically, this is a question about the interplay between three formal power series. First we might as well say our class of rooted trees is simple and so is defined by a formal power series. Second we have the hook weight sequence which we can view as the coefficient sequence of a formal power series. Finally we have the series expansion of the Green function resulting from combining these. Any two of these series determines the third and so the main theorem of this area, independently found in pure combinatorics [4], in B-series [5], and in quantum field theory [6] gives this relationship explicitly in the favoured language of each community (via coefficient extraction, via a differential equation, or via the universal property). Making explicit the connections between these viewpoints is the main subject of Jones' work in [1, 7].

To state the main theorem we need a little more notation. Given a hook weight sequence B we already defined L_B via $L_B(x^n) = B_{n+1}x^{n+1}$. Also define a shifted version L_B^* via $L_B^*(x^n) = (n + 1)B_{n+1}x^n$ and define θ to be the operator that takes $f(z)$ to $\theta(f)(z) = z\frac{d}{dz}f(z)$. Combinatorially θ is the pointing operator. Then we get the following theorem.

Theorem 3 (Theorem 3.1 of [1]) *Let B be a hook weight sequence with associated hook weight Feynman rules ϕ. Let $\mathcal{T}$ be a simple tree class defined by $T(x) = xB_+(f(T(x)))$. Let $G(x) = \sum_{t \in T} \phi(t)x^{|t|}$. Then*

1. *G satisfies the recurrence*

$$[x^n]G(x) = B_n[x^{n-1}]f(G(x)), \quad \forall k \geq 1.$$

2. *G is a solution to the differential equation:*

$$G'(x) = L_B^*(1 + \theta)(f(G(x))).$$

3. *G satisfies*

$$G(x) = L_B(f(G(x))).$$

What is nice about recognizing that different communities have been studying Dyson-Schwinger equations with this special class of Feynman rules is that we can then try to use the insights of one community to get something new for another community. The pure combinatorics community had built up a large family of examples, some rather exotic, which can be imported to give leading log toy models for different combinatorial Dyson-Schwinger equations, with various hook weight Feynman rules, all of which have nice Green functions. See [7] for a table with a large collection of these examples. In the other direction, the differential equation form of the main theorem was known in the pure combinatorics community in special cases but its full power was not appreciated, so we can take this idea to obtain new techniques for finding new examples, see [1, Table 1].

To conclude this chapter here is an example of a nice Green function coming from this framework. This example was known[1] but to my knowledge did not appear in print before [1] so it is fair to call it underappreciated. The combinatorial Dyson-Schwinger equation in this case is

$$T(x) = \mathbb{I} - x B_+ \left(\frac{1}{T(x)^2} \right).$$

and we use the tree factorial Feynman rules. Then Theorem 3 tells us that the Green function is

$$G(z) = (1 - 3z)^{\frac{1}{3}},$$

which is a nice closed form.

Dozens more examples of varying levels of complexity and physicality can be read off the tables of [7].

References

1. Jones, B.R., Yeats, K.: Tree hook length formulae, Feynman rules and B-series. Ann. Inst. Henri Poincaré D. **2**(4), 413–430 (2015). arXiv:1412.6053
2. Postnikov, A.: Permutohedra, associahedra, and beyond. Int. Math. Res. Not. **6**, 1026–1106 (2009)
3. Kuba, M., Panholzer, A.: Bilabelled increasing trees and hook-length formulas. Eur. J. Combin. **33**(2), 248–258 (2012)
4. Kuba, M., Panholzer, A.: A unifying approach for proving hook-length formulas for weighted tree families. Graphs Combin. **29**(6), 1839–1865 (2013)
5. Mazza, C.: Simply generated trees, b-series and wigner processess. Random Struct. Algorithms **25**(3), 293–310 (2004)
6. Panzer, E.: Hopf-algebraic renormalization of Kreimer's toy model. Master's thesis, Humboldt-Universität zu Berlin (2011)
7. Jones, B.R.: On tree hook length formulae, Feynman rules and B-series. Master's thesis, Simon Fraser University (2014)

---

[1]Dirk Kreimer personal communication.

Chapter 9
Chord Diagram Expansions

9.1 Converting the Dyson-Schwinger Equation to Differential Form

The next step on our path following Dyson-Schwinger equations from combinatorics to physics is to allow richer Feynman rules, rich enough to capture physically relevant situations. We will, however, be sticking to the single scale case as one sees in propagator insertions. We still want to have combinatorial control over the answer and so the first step is to rewrite the analytic Dyson-Schwinger equation so as to unwind the analytic side from the combinatorial side as much as possible.

Let's see how this rewriting works in an example. This is Example 3.7 from [1, 2]. Recall the Yukawa example of [3] as discussed in Chap. 6. We had the analytic Dyson-Schwinger equation

$$G(x, L) = 1 - \frac{x}{q^2} \int d^4k \frac{k \cdot q}{k^2 G(x, \log(k^2/\mu^2))(k + q)^2} - \cdots \Big|_{q^2=\mu^2}$$

where $L = \log(q^2/\mu^2)$.

Now we will make the following transformations: expand G in its second argument, convert logarithms to powers using $\frac{d^k y^\rho}{d\rho^k}\big|_{\rho=0} = \log^k(y)$, swap the order of the operators and recombine the expansion of G. Specifically, substitute the Ansatz

$$G(x, L) = 1 - \sum_{k \geq 1} \gamma_k(x) L^k$$

into the analytic Dyson-Schwinger equation to get

© The Author(s) 2017
K. Yeats, *A Combinatorial Perspective on Quantum Field Theory*,
SpringerBriefs in Mathematical Physics 15, DOI 10.1007/978-3-319-47551-6_9

$$\sum_{k\geq 1} \gamma_k(x) L^k$$

$$= \frac{x}{q^2} \int d^4k \sum_{\ell_1+\cdots+\ell_s=\ell} \frac{(k\cdot q)\gamma_{\ell_1}(x)\cdots\gamma_{\ell_s}(x)\log^\ell(k^2/\mu^2)}{k^2(k+q)^2} - \cdots \Bigg|_{q^2=\mu^2}$$

$$= \frac{x}{q^2} \sum_{\ell_1+\cdots+\ell_s=\ell} \gamma_{\ell_1}(x)\cdots\gamma_{\ell_s}(x) \int d^4k \frac{(k\cdot q)\log^\ell(k^2/\mu^2)}{k^2(k+q)^2} - \cdots \Bigg|_{q^2=\mu^2}$$

$$= \frac{x}{q^2} \sum_{\ell_1+\cdots+\ell_s=\ell} \gamma_{\ell_1}(x)\cdots\gamma_{\ell_s}(x) \int d^4k \frac{(k\cdot q)(-1)^\ell (d/d\rho)^\ell (k^2/\mu^2)^{-\rho}|_{\rho=0}}{k^2(k+q)^2} - \cdots \Bigg|_{q^2=\mu^2}$$

$$= \frac{x}{q^2} \sum_{\ell_1+\cdots+\ell_s=\ell} \gamma_{\ell_1}(x)\cdots\gamma_{\ell_s}(x)(-1)^\ell$$

$$\cdot (d/d\rho)^\ell (\mu^2)^\rho \int d^4k \frac{k\cdot q}{(k^2)^{1+\rho}(k+q)^2} - \cdots \Bigg|_{q^2=\mu^2}\Bigg|_{\rho=0}$$

$$= x\left(1 - \sum_{k\geq 1} \gamma_k(x)\left(\frac{d}{d(-\rho)}\right)^k\right)^{-1} \frac{(\mu^2)^\rho}{q^2} \int d^4k \frac{k\cdot q}{(k^2)^{1+\rho}(k+q)^2} - \cdots \Bigg|_{q^2=\mu^2}\Bigg|_{\rho=0}$$

$$= x\left(1 - \sum_{k\geq 1} \gamma_k(x)\left(\frac{d}{d(-\rho)}\right)^k\right)^{-1} \frac{(\mu^2)^\rho}{(q^2)^\rho} \int d^4k_0 \frac{k_0\cdot q_0}{(k_0^2)^{1+\rho}(k_0+q_0)^2} - \cdots \Bigg|_{q^2=\mu^2}\Bigg|_{\rho=0}$$

where $q = rq_0$ with $r\in\mathbb{R}, r^2=q^2, q_0^2=1$ and $k=rk_0$

$$= x\left(1 - \sum_{k\geq 1} \gamma_k(x)\left(\frac{d}{d(-\rho)}\right)^k\right)^{-1} (e^{-L\rho}-1)F(\rho)\Bigg|_{\rho=0}$$

where

$$F(\rho) = \frac{1}{q^2} \int d^4k \frac{k\cdot q}{(k^2)^{1+\rho}(k+q)^2}\Bigg|_{q^2=1}.$$

Then recombine to get

$$G(x,L) = 1 - xG\left(x, \frac{d}{d(-\rho)}\right)^{-1} (e^{-L\rho}-1)F(\rho)\Bigg|_{\rho=0}.$$

$F(\rho)$ is the Feynman integral of the primitive evaluated at $q^2=1$ with the propagator we are inserting into regularized. Call $F(\rho)$ the **Mellin transform** of the primitive. One nice thing about working with Dyson-Schwinger equations renormalized using subtraction is that this transformation makes the regularization appear naturally rather than having regularization be an artificial step added in by hand.

Now we would like to be able to do this kind of transformation for any Dyson-Schwinger equation. In particular, given the single equation combinatorial Dyson-Schwinger equation

$$T(x) = \mathbb{I} - \sum_{k\geq 1} x^k B_+^k(T(x)Q(x)^k)$$

where $Q(x) = T(x)^{-s}$ (compare Sect. 5.4), we want to work with

$$G(x, L) = 1 - \sum_{k \geq 1} x^k G\left(x, \frac{d}{d(-\rho)}\right)^{1-sk} (e^{-L\rho} - 1) F_k(\rho)\bigg|_{\rho=0} \qquad (9.1)$$

where s is a positive integer and the $F_k(\rho)$ are the Mellin transforms of the primitives. As written this doesn't necessarily make analytic sense at all since $G\left(x, \frac{d}{d(-\rho)}\right)$ is a priori only a pseudodifferential operator. Thinking formally this is no problem—it is a valid series operator. Having things work out analytically involves both guaranteeing the series in $d/d\rho$ makes sense and that operators can be commuted as needed in the argument, which requires some analytic hypotheses. The necessary analytic conditions are given in [4]. We will just assume that our Feynman rules are sufficiently well behaved and so work with (9.1).

This transformation is the first step towards the more substantial reductions of [1, 2] which have been used in [5] and [6] to consider QED and QCD showing a possible way to avoid the Landau pole and used by Marc Bellon and collaborators [7–9] to investigate the Wess-Zumino model.

The goal of the rest of this chapter is to give a series solution to (9.1) which is indexed by rooted connected chord diagrams.

9.2 Rooted Connected Chord Diagrams

A *rooted chord diagram* of size n is simply a matching of $\{1, 2, \ldots, 2n\}$, that is, a decomposition of $\{1, 2, \ldots, 2n\}$ into disjoint pairs called *chords*. We can draw a rooted chord diagram on a circle, hence the name chord diagram. See Fig. 9.1 for two examples of size 4. The marked vertex, also known as the root vertex, corresponds to 1 and the other vertices are numbered in counterclockwise order, so the first chord diagram of Fig. 9.1 corresponds to the matching $(1, 4), (2, 7), (3, 5), (6, 8)$.

Given a chord diagram we say two chords, c_1 and c_2, *cross* if $c_1 = \{v_1 < v_2\}$, $c_2 = \{w_1 < w_2\}$ gives $v_1 < w_1 < v_2 < w_2$ or $w_1 < v_1 < w_2 < v_2$. This corresponds to the intuitive notion of crossing in the circle drawing of the chord diagram.

Given a chord diagram we can form the *oriented intersection graph* of the chord diagram as follows. The vertices of this graph are the chords of chord diagram. There is an edge from $c_1 = \{v_1 < v_2\}$ to $c_2 = \{w_1 < w_2\}$ if $v_1 < w_1 < v_2 < w_2$, that is

Fig. 9.1 Two rooted chord diagrams

Fig. 9.2 The terminal
chords and intersection order
of a chord diagram

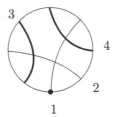

if c_1 and c_2 cross and c_1 has a smaller first endpoint. A chord diagram is ***connected***
if its intersection graph is connected. For example the first chord diagram in Fig. 9.1
is connected but the second one is not because the chords can be separated into
two subsets with no crossings between the subsets. A ***connected component*** of a
chord diagram is a subset of chords corresponding to a connected component of the
intersection graph or equivalently a subset of chords which is itself connected but
with no crossings with other chords of the diagram.

Rooted chord diagrams and rooted connected chord diagrams have been studied by
the combinatorics community, see [10, 11]. The particular parameters we need, while
rich and useful for quantum field theory, are unexpected from a pure combinatorics
perspective and so do not have a long history of prior study.

The first of these more unusual aspects of chord diagrams is the notion of a
terminal chord. We say a chord is ***terminal*** if its vertex in the oriented intersection
graph of the chord diagram has no outgoing edges. Terminal chords do not cross any
later chords. For example in Fig. 9.2 the terminal chords are indicated with the fat
edges.

We also need to specify an indexing for the chords. Chords do inherit an indexing
from the counterclockwise order (say by their first appearing endpoint). This is not
the indexing we want.

Definition 19 Let C be a rooted connected chord diagram. The ***intersection order***
of the chords of C is defined as follows.

- The root chord of C is the first chord in the intersection order.
- Remove the root chord of C and let $C_1, C_2, \ldots, C_n$ be the connected components
  of the result ordered by their first vertex in counterclockwise order.
- Inductively order the chords of each C_i by the intersection order. The intersection
  order of the chords of C is first the root, then the chords of C_1 in intersection order,
  then the chords of C_2 in intersection order, and so on.

For example the intersection order is indicated in the example in Fig. 9.2. Note
that in this case the intersection order does not match the order given by the first
endpoints of the chords. However, the terminal chords are still those which do not
cross any later chords. Another way to see the terminal chords is that they correspond
to the base case of Definition 19. We will write $b(C)$ to denote the index of the ***first
terminal chord*** in intersection order.

Later on we will want to *decorate* chord diagrams by associating a nonnegative integer to each chord. The *size* of a decorated chord diagram is the sum of its decorations. Thus the undecorated case corresponds to the case where all decorations are 1. There is another trickier invariant which we'll need in the decorated case and which is discussed in Sect. 9.4.

9.3 The $s = 2, k = 1$ Result

The first case of (9.1) to consider is the case where $s = 2$ and there is a single primitive ($k = 1$). This corresponds to the combinatorial Dyson-Schwinger equation

$$T(x) = \mathbb{I} - x B_+ \left(\frac{1}{T(x)} \right),$$

that is, to plane rooted trees. A physical example with these combinatorics is the Yukawa example studied by Broadhurst and Kreimer (see Sect. 9.1). In general it corresponds to inserting a single 1-loop propagator correction into itself in one place recursively in all possible ways (nested and chained in that one place). At the analytic level this means we are trying to solve (9.1) with $s = 2$ and $k = 1$. Specifically we are trying to solve

$$G(x, L) = 1 - xG \left(x, \frac{d}{d(-\rho)} \right)^{-1} (e^{-L\rho} - 1)F(\rho) \Big|_{\rho=0}.$$

Given a rooted connected chord diagram C, let the indices of its terminal chords in intersection order be $b(C) = t_1 < t_2 < \cdots < t_k$. The main result of [12], work with Nicolas Marie, is that the series solution to this equation can be written

$$G(x, L) = 1 - \sum_{i \geq 1} \frac{(-L)^i}{i!} \sum_{\substack{C \\ b(C) \geq i}} x^{|C|} f_{b(C)-i} f_0^{|C|-k} \prod_{j=2}^{k} f_{t_j - t_{j-1}} \qquad (9.2)$$

where the sum is over rooted connected chord diagrams with the indicated restriction and

$$F(\rho) = \frac{f_0}{\rho} + f_1 + f_2\rho + \cdots .$$

Note that k and the t_j depend on C but this has been suppressed in the notation.

This is an interesting result because it gives the Green function as a combinatorial expansion over chord diagrams. Of course, we began with the Green function as an expansion over Feynman graphs, but the Feynman diagram expansion isn't a fully combinatorial expansion because each graph is weighted by its Feynman integral which is a complicated analytic object. The chord diagram expansion is different because for a fixed power of L and fixed power of x each chord diagram contributes a single monomial in the coefficients of the expansion of $F(\rho)$.

This result is also interesting because it shows explicitly how the expansions in x for fixed powers of L relate to each other. Specifically, as the power of L changes two aspects of the chord diagram expansion in L change. First, as the power of L increases more diagrams are truncated from the sum; in particular we can easily see the standard fact that the exponent of x must be at least as large as the exponent of L. Second, one of the f_ℓ in the monomial shifts depending on the power of L, namely $f_{b(C)-i}$ for L^i.

The result is useful because a combinatorial understanding of chord diagrams gives us a better understanding of the Green function. Perhaps most interesting is to look at asymptotics of chord diagrams and what it can give us. This has been pursued with Julien Courtiel in [13], one important aspect of which will be discussed in Chap. 10.

Another interesting facet of this result is that now that we have an explicit combinatorial expansion for the Green function, we can look at other equations for the Green function from quantum field theory and see what they look like as combinatorial expressions for chord diagrams. Most striking along these lines is the renormalization group equation which ends up being a key part of the proof of (9.2).

The *renormalization group equation* tells us how the x and L derivatives of the Green function relate. That is, it tells us how the Green function changes as the energy scale changes. In the most naive possible situation the Green function would simply be invariant under change of scale, but this is too simple for the kinds of theories we are interested in. Rather, two other things can happen. First, the coupling constant may change with the energy scale; this makes sense because we are working with the renormalized coupling not the bare coupling. The *beta function* of the theory tells us how the coupling constant changes with the energy scale. The dynamics of the beta function are extremely important in a qualitative understanding of quantum field theories. The other thing that can happen is that the field gets rescaled. This gives the *anomalous dimension* of the field, and we will denote it by γ.

In our context the renormalization group equation reads

$$\left(\frac{d}{dL} + \beta(x)\frac{d}{dx} + 2\gamma(x)\right) G(x, L) = 0$$

where $2\gamma(x)$ is the coefficient of L in $G(x, L)$ and in this case $\beta(x) = -2x\gamma(x)$. In general $\beta(x)$ is an appropriate linear combination of the anomalous dimensions of the theory, but in this case we began with a single equation, not a system of Dyson-Schwinger equations, so there is only the one anomalous dimension $\gamma(x)$.

Applying the renormalization group equation to the chord diagram expansion and equating coefficients gives the recurrence on chord diagrams

$$g_{k,i} = \sum_{\ell=1}^{i-1}(2\ell - 1)g_{1,i-\ell}g_{k-1,\ell} \qquad \text{for } 2 \le k \le i \qquad (9.3)$$

where

$$g_{k,i} = \sum_{\substack{C \\ |C|=i \\ b(C) \geq i}} f_{b(C)-i} f_0^{|C|-k} \prod_{j=2}^{k} f_{t_j - t_{j-1}}.$$

To illustrate the idea of the proof of this recurrence without getting bogged down in details, let's forget the monomials in the f_ℓ for the moment. Let s_n be the number of rooted connected chord diagrams with n chords. Then the recurrence becomes

$$s_i = \sum_{\ell=1}^{i-1} (2\ell - 1) s_{i-\ell} s_\ell \qquad \text{for } i \geq 2$$

This is a classical recurrence originally due to Stein [14] and rephrased by Riordan. Nijenhuis and Wilf [10] prove this recurrence using a decomposition of chord diagrams.

Given a rooted connected chord diagram C, define the ***root share decomposition*** of C to be the rooted connected chord diagrams C_1 and C_2 obtained as follows. First remove the root of C. Of the connected components which remain, the one rooted at the next vertex counterclockwise from the original root is C_2. C_1 is the rooted connected chord diagram resulting from removing the chords of C_2 from C. See Fig. 9.3 for an example. The root of C_2 is marked with an $\times$ in the example.

This process can be reversed, but we need one extra piece of information. Given C_1 and C_2 first choose $1 \leq k \leq |C_2| - 1$. Each k corresponds to an interval along the circle between successive vertices of C_2, excluding the interval between the last vertex and the root vertex. We can then insert C_1 into C_2 in interval k by putting the root of C_1 immediately before the root of C_2 and putting all other vertices of C_1 in interval k. See Fig. 9.3.

Insertion and the root share decomposition are inverses of each other and so give a bijection between rooted connected chord diagrams and triples of two rooted connected chord diagrams C_1 and C_2 and an integer $1 \leq k \leq |C_2| - 1$. Therefore these two sets are equinumerous which gives Stein's recurrence. Following through this proof keeping track of the terminal chords gives (9.3). See Proposition 4.1 of [12] for details.

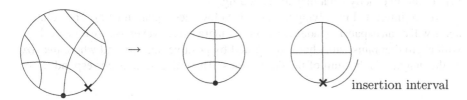

Fig. 9.3 The root share decomposition

The proof of (9.2) in [12] is by recurrences. We show the Green function defined as the solution to the Dyson-Schwinger equation and the chord diagram expansion satisfy the same recurrence and have the same initial conditions. This is done in two parts. The first part is the renormalization group equation. We know from quantum field theory that the solution to the Dyson-Schwinger equation satisfies the renormalization group equation while the argument sketched above shows that the chord diagram expansion also satisfies it. This tells us that the coefficients of the powers of L (which are series in x) satisfy the same recurrence in both cases. It remains to show that the coefficient of L agrees between the solution to the Dyson-Schwinger equation and the chord diagram expansion. That is, it remains then to show that the anomalous dimension matches its chord diagram expansion.

Showing this requires a second recurrence to show the coefficients in x agree. This recurrence is considerably more technical and involves going through another class of combinatorial objects, certain leaf-labelled plane binary trees. For the purposes of [12] these binary trees were purely a technical device, but subsequent work, discussed in the next section, showed they do carry some important structure.

9.4 Binary Trees and the General Result

To generalize the chord diagram expansion to a wider class of Dyson-Schwinger equations we need to define a map from rooted connected chord diagrams to plane binary trees. The leaves of the trees will be in bijection with the chords of the diagram.

The map is as follows. Associate the chord diagram with only one chord to the binary tree with only one vertex; the single chord corresponding to this one vertex. Given a rooted connected chord diagram C with $|C| > 1$, let its root share decomposition be C_1 and C_2 inserted into interval k. Recursively form the binary trees of C_1 and C_2. Call them T_1 and T_2. Find the kth vertex of T_2 in a pre-order traversal[1]; call this vertex v. Take T_2, remove the subtree rooted at v putting a new vertex w where v used to be. Let T_1 be the right subtree of w and let the subtree originally rooted at v be the left subtree of w. This defines the tree of C. See Fig. 9.4 for a schematic of the tree insertion and Fig. 9.5 for an example. The appendix of [12] has many examples.

Note that these trees are not insertion trees for Feynman graphs. They are representing a different tree-like structure of the chord diagrams. In fact they are a bit mysterious in exactly what they are encoding.

We are interested in solving the Dyson-Schwinger equation (9.1). The solution again will be an expansion indexed by chord diagrams. However, we are now working with chord diagrams with chords decorated by positive integers and where the size of the diagram is the sum of the decorations. These decorations correspond to the

---

[1] In a pre-order traversal the root vertex comes first. Next come all the vertices of the subtree rooted at the left child of the root ordered following a pre-order traversal. Finally come all the vertices of the subtree rooted at the right child of the root also ordered following a pre-order traversal.

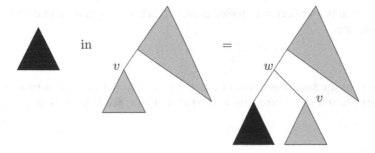

Fig. 9.4 A schematic of the tree insertion

Fig. 9.5 A rooted connected chord diagram and its binary tree

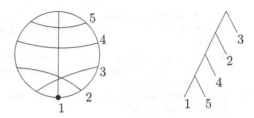

different loop orders of the primitives. Given a chord diagram C and a chord c of C write d_c for the decoration of c.

We also need a weight given by the binary tree. Specifically, given a rooted connected chord diagram C and a chord c of C, build the tree of C. Begin at the leaf associated to c and walk up and left from c as far as possible. Denote the length of this path by $v(c)$. For example in Fig. 9.5 $v(1) = 0$ and $v(4) = 1$. Then the overall weight of C, depending on the parameter s, is

$$w(C) = \prod_{c \in C} \binom{d_c s + v(c) - 2}{v(c)}$$

where the product is over all chords of C.

Given a decorated rooted connected chord diagram C, write $|C|$ for the sum of the decorations, let the indices of its terminal chords in intersection order be $b(C) = t_1 < t_2 < \cdots < t_k$ and write $\mathrm{ter}(C) = \{t_1, t_2, \ldots, t_k\}$. Then the main result of [15] is that the solution to (9.1) is

$$G(x, L) = 1 - \sum_{\substack{i \geq 1}} \frac{(-L)^i}{i!} \sum_{\substack{C \\ b(C) \geq i}} x^{|C|} w(C) f_{d_{b(C)}, b(C) - i} \prod_{j=2}^{k} f_{d_{t_j}, t_j - t_{j-1}} \prod_{c \neq \mathrm{ter}(C)} f_{d_c, 0}$$

where the sum is over decorated rooted connected chord diagrams with the indicated restrictions and

$$F_k(\rho) = \frac{f_{k,0}}{\rho} + f_{k,1} + f_{k,2}\rho + \cdots.$$

The proof has the same basic structure as the $s = 2$, $k = 1$ case but is substantially more intricate and was worked out with Markus Hihn. See [15] for details.

References

1. Yeats, K.: Rearranging Dyson-Schwinger equations. Mem. Am. Math. Soc. **211** (2011)
2. Yeats, K.A.: Growth estimates for Dyson-Schwinger equations. Ph.D. thesis, Boston University (2008)
3. Broadhurst, D., Kreimer, D.: Exact solutions of Dyson-Schwinger equations for iterated one-loop integrals and propagator-coupling duality. Nucl. Phys. B **600**, 403–422 (2001). arXiv:hep-th/0012146
4. Marie, N.: On Laplace Borel resummation of Dyson Schwinger equations. Ph.D. thesis, Simon Fraser University (2014)
5. van Baalen, G., Kreimer, D., Uminsky, D., Yeats, K.: The QED beta-function from global solutions to Dyson-Schwinger equations. Ann. Phys. **234**(1), 205–219 (2008). arXiv:0805.0826
6. van Baalen, G., Kreimer, D., Uminsky, D., Yeats, K.: The QCD beta-function from global solutions to Dyson-Schwinger equations. Ann. Phys. **325**(2), 300–324 (2010). arXiv:0805.0826
7. Bellon, M.: Approximate differential equations for renormalization group functions in models free of vertex divergencies. Nucl. Phys. B **826**, 522–531 (2010). arXiv:0907.2296
8. Bellon, M.: An efficient method for the solution of Schwinger-Dyson equations for propagators. Lett. Math. Phys. **94**(1), 77–86 (2010). arXiv:1005.0196
9. Bellon, M.P., Schaposnik, F.A.: Higher loop corrections to a Schwinger–Dyson equation. Lett. Math. Phys. **103**(8), 881–893 (2013). arXiv:1205.0022
10. Nijenhuis, A., Wilf, H.S.: The enumeration of conected graphs and linked diagrams. J. Combin. Theory A **27**, 356–359 (1979)
11. Flajolet, P., Noy, M.: Analytic combinatorics of chord diagrams. In: In Formal Power Series and Algebraic Combinatorics, pp. 191–201 (2000)
12. Marie, N., Yeats, K.: A chord diagram expansion coming from some Dyson-Schwinger equations. Commun. Number Theory Phys. **7**(2), 251–291 (2013). arXiv:1210.5457
13. Courtiel, J., Yeats, K.: Terminal chords in connected chord diagrams. arXiv:1603.08596
14. Stein, P.R.: On a class of linked diagrams I. Enumer. J. Combin. Theory A **24**, 357–366 (1978)
15. Hihn, M., Yeats, K.: Generalized chord diagram expansions of Dyson-Schwinger equations. arXiv:1602.02550

Chapter 10
Differential Equations and the (Next-To)^m Leading Log Expansion

10.1 The (Next-To)^m Leading Log Expansions

We are still one important step away from a physical understanding of solutions to Dyson-Schwinger equations because we are still working with series expansions but we want functions. Furthermore, we know the number of rooted chord diagrams is counted by double factorial, specifically the number of rooted chord diagrams with n chords is $(2n - 1)!! = (2n - 1)(2n - 3)(2n - 5) \cdots 1$. Imposing connectivity does not change the fundamentally factorial growth. Thus, as expected for quantum field theory, our series are divergent. In principal this means that the expansion is only telling us about the function in an infinitesimal neighbourhood of the expansion point. In practice, however, things seem to be much better. Summing the perturbative expansion as far as it has been computed is highly predictive in quantum field theory. Mathematically, this is hinting that the series which occur in the perturbative expansion have extra structure.

Resumming divergent series is a subtle business. Much has been done both theoretically and practically, see the references in [1], but much remains to be understood. These ideas are beyond the scope of this brief and not particularly combinatorial in flavour and so will not be explored further here.

What then can we hope to do? First we can ask about asymptotics for the coefficients of our expansions. Since our series are essentially generating functions we can apply the tools of asymptotic combinatorics. Some results along these lines were obtained for the chord diagram expansion in the $s = 2, k = 1$ case in [2]. Asymptotics are necessary for any qualitative understanding of the behaviour after resummation.

Another thing we can do to extract functions of physical significance out of the perturbative expansion is to think again about how the expansion is indexed and use that to break it up in a different way. We have triangular sums of the form

$$G(x, L) = 1 + \sum_{i \geq 1} \sum_{j \geq i} a_{i,j} L^i x^j.$$

© The Author(s) 2017
K. Yeats, *A Combinatorial Perspective on Quantum Field Theory*,
SpringerBriefs in Mathematical Physics 15, DOI 10.1007/978-3-319-47551-6_10

Rather than thinking of the sum first as an expansion in one of the variables with coefficients which are series in the other variable, we can write the expansion as

$$G(x, L) = \sum_{m \geq 0} \sum_{i \geq 0} a_{i,i+m} x^m (Lx)^i.$$

If there are further parameters such as masses and scattering angles they can be included in the $a_{i,i+m}$. The $m = 0$ part of this sum, namely the terms of $G(x, L)$ where the powers of L and x are the same, is known as the *leading log expansion*. The $m = 1$ part of this sum, namely the terms of $G(x, L)$ where the power of x is one more than the power of L is known as the *next-to-leading log expansion*. The $m = 2$ part is known as the *next-to-next-to-leading log expansion* and so on.

This leading log language comes from the fact that L is $\log(q^2/\mu^2)$ or some similar logarithm of an energy scale, while x is the coupling constant. So the leading log expansion captures the maximal powers of x relative to the powers of the energy scale, and so is in an important sense the leading term. The next-to-leading log expansion is the part suppressed by one power of x, and so on.

The expansions of this log hierarchy are simpler analytically and combinatorially and so are a place where we can hope to obtain and understand the functions, not just the expansions.

10.2   Combinatorial Expansions of the Log Expansions

There are two recent combinatorial approaches to understanding the next-to^m leading log expansions. The first of these is by Krüger and Kreimer [3]. The second follows from the chord diagram expansion of the previous chapter with a combinatorial analysis by Courtiel [2].

To illustrate both approaches consider first the Dyson-Schwinger equation (9.1) with $s = 2$ but for now allow any k. An example of this would be fermion propagator corrections in Yukawa theory. The leading log expansion is easy enough to understand. Primitives with more than 1 loop are suppressed by at least one power of x and so do not appear in the leading log expansion. This leaves us in the $s = 2$, $k = 1$ case, but even more is stripped away in the leading log expansion as we only take the leading part of each insertion. The question is how to capture this leading part in a generalizable way.

Krüger and Kreimer's answer is to map the renormalization Hopf algebra of Feynman graphs to a Hopf algebra of words based on a quasishuffle-deconcatenation Hopf algebra. This is like the shuffle-deconcatenation Hopf algebra of Sect. 4.1 except that the shuffle is a quasishuffle, that is extra combination terms appear in the shuffle formula. The B_+s in this word algebra correspond to prepending a letter. The letters initially correspond to the primitive Feynman graphs—for the leading log case we have just one. The key here is that an arbitrary word cannot be written in terms of just the quasishuffle product, so Krüger and Kreimer rewrite it as a linear combination of pieces made with the quasishuffle, a concatenation commutator, and letters which are either the basic letters corresponding to primitives or are new letters associated to the extra term in the quasishuffle. The pure quasishuffle part in the 1 loop primitive corresponds to the leading log part while any parts with commutators or other letters show up suppressed by powers of x. Mapping the Dyson-Schwinger equation over to this word Hopf algebra gives an easy to solve differential equation for this pure quasishuffle part.

Courtiel and I instead use the chord diagram expansion to understand how to take only the leading log part of each insertion. From (9.2) we want only the $|C| = i$ part; that is we only want the $b(C) = |C|$ part:

$$\sum_{b(C)=|C|} \frac{(-Lxf_0)^{|C|}}{|C|!}$$

which is the exponential generating function for rooted connected chord diagrams with $b(C) = |C|$. These are easy to count from the root-share decomposition (which was defined in Sect. 9.3). Every component after removing the root contributes at least one terminal chord, so the C_1 of the root share decomposition must be simply the root chord and there are $2|C| - 3$ ways to do this. So recursively there are $(2|C| - 3)!!$ rooted connected chord diagrams with $b(C) = |C|$.

With either technique the $s = 2$ leading log expansion is

$$1 - \sqrt{1 - 2Lxf_0}.$$

Next consider how this generalizes. For Krüger and Kreimer, as they go further down the hierarchy of log expansions they need to consider quasishuffles involving higher loop primitives, commutators, and new letters from the quasishuffle. For any fixed inventory of these exotic pieces, shuffling on more copies of the 1 loop primitive does not change the amount of x suppression, so they collect together the contributions into families based on this inventory. For any given family, the Dyson-Schwinger equation mapped to words gives a differential equation. For a fixed m only finitely many families contribute. Making the argument for a general family they get a master differential equation. Some coefficients in the master differential equation come from counting how many ways certain matrices can be built. In any given case these values can be calculated but they are not well understood in general. They work out the answers explicitly in the Yukawa ($s = 2$) and QED($s = 1$) cases down to the next-to-next-to leading log expansion.

Courtiel and I have looked at the $s = 2$, $k = 1$ case in [2]. For the next-to^m leading log expansion we need to consider all distributions of terminal chords which give rooted connected chord diagrams with $b(C) \leq |C| - m$. We work them out explicitly for the next-to and next-to-next-to leading log expansions. The root-share decomposition gives recurrences hence differential equations in each case. This is our analogue of the master equation. Furthermore, we prove that the diagrams with precisely the last m chords terminal dominate and so, asymptotically, the coefficients of any of the log expansions in the $s = 2$, $k = 1$ case are dominated by the contributions of f_0 and f_1. See Sect. 3 of [2] for precise results.

On the common domain of applicability both groups' results are the same. This had to happen because we are both describing the same underlying physics. Both groups have a combinatorial perspective and a combinatorially derived master equation. However, the basic objects in each case are quite different. It is not clear if or how the objects could be directly connected combinatorially, but it is tantalizing to think that these different methods hint at a deep connection between words and chord diagrams. Also the groups index the new numbers appearing differently. For Courtiel and I they come from subsequent terms in the expansion for the primitive. For Krüger and Kreimer they come from letters, both the original letters from primitives and those from the quasishuffle term which are identified noncanonically with other Feynman graphs. Krüger and Kreimer's techniques immediately hold at a high level of generality. Courtiel and mine will take some work to generalize.

Ultimately the different strengths and perspectives of the two approaches are enriching and exciting.

References

1. Klaczynsky, L.: Resurgent transseries & Dyson-Schwinger equations. arXiv:1601.04140
2. Courtiel, J., Yeats, K.: Terminal chords in connected chord diagrams. arXiv:1603.08596
3. Krüger, O., Kreimer, D.: Filtrations in Dyson-Schwinger equations: next-to^j -leading log expansions systematically. arXiv:1412.1657

Part III
Feynman Periods

Chapter 11
Feynman Integrals and Feynman Periods

In the chord diagram expansion and the Krüger-Kreimer log expansion we saw that the primitive Feynman diagrams were the analytic input. In both cases, with these primitives taken as black boxes, there was a nice combinatorial understanding of how to put things together. Now it is time to look at these black boxes.

In general this is very hard. There is a whole community with expertise in calculating diagrams, see for example the articles from [1]. To keep things simple, here we will stick to primitive 4-point Feynman graphs in ϕ^4. In graph theory language this means that we are looking at graphs which come from a 4-regular graph (hence in ϕ^4 theory, see Sect. 5.2) with one vertex removed (hence 4 external edges, that is, 4-point). Next we need to understand what primitive means graph theoretically. Recall from Sect. 5.2 the notion of superficial degree of divergence. In the case of a ϕ^4 graph γ this says that γ is divergent if $4\ell(\gamma) - 2|E(\gamma)| \geq 0$ where $\ell(\gamma)$ is the loop number of the graph and $E(\gamma)$ is the set of internal edges. By Euler's formula this inequality is equivalent to the statement that γ has 4 or fewer external edges. Note in particular that double edges are divergent. So let G be the result of removing a vertex from a 4-regular graph K and let $\gamma \subsetneq G$ be divergent. We can also view γ as a subgraph of K. But γ has 4 or fewer external edges and K has none, so there must be an edge cut of 4 edges or fewer which separates γ from the rest of K. Conversely, any 4 edge cut of K other than the ones which cut off only a single vertex gives a divergence subgraph. Since all vertices of K have even degree if K has no such 4 edge cuts then it also has no such 5 edge cuts. The property of having no 5 edge cuts other than the ones which cut off a single vertex is called being ***internally 6-edge-connected***.

For the remainder of this brief we will be primarily concerned with these primitive 4-point ϕ^4 graphs. Furthermore we will look only at the massless case. This case is already very interesting both combinatorially and number theoretically [2–9]. Furthermore, many of the techniques generalize to the massive case and to more realistic physical theories [10, 11] as well as, on the combinatorial side, to not necessarily 4-regular graphs, see for example [12].

© The Author(s) 2017
K. Yeats, *A Combinatorial Perspective on Quantum Field Theory*,
SpringerBriefs in Mathematical Physics 15, DOI 10.1007/978-3-319-47551-6_11

In Sect. 5.6 we saw the Feynman rules in momentum space for ϕ^4 theory. For the purposes of Feynman periods the Feynman integral in parametric space is very important, and position space is also useful.

We first need a few observations about the momentum space integrals. In building the momentum space integrand we put an arbitrary orientation on the graph and then assigned a momentum to each edge with the restriction that the sum of the momenta coming into any vertex is the same as the sum of the momenta going out of that vertex. Graph theoretically this means that the momenta give a *flow* on the graph with values in $\mathbb{R}^4$. For the purposes of the Feynman period we can set all the external momenta to 0.

Recall the cycle space of a graph from Sect. 5.5. As observed in Sect. 5.6 the dimension of the cycle space is the number of free momentum variables in the integrand. To be explicit, if we choose a set of oriented cycles $C_1, C_2, \ldots, C_\ell$ which is a basis for the cycle space, then we can let the momentum variables be $p_1, \ldots, p_\ell$ and then take the momentum of edge e to be $p_e = \sum \pm p_i$ where p_i appears with a positive sign if cycle C_i goes through edge e in the direction of e and with a negative sign if it goes through in the opposite direction.

Proposition 5 *Let G be a primitive 4-point graph in ϕ^4 with loop number ℓ. The following three integrals converge and give the same number.*

1. *(momentum space) Choose a unit vector in $\mathbb{R}^4$ and call it 1. Choose a basis $C_1, C_2, \ldots, C_\ell$ of the cycle space and assign momenta to the edges as described above. Set $p_\ell = 1$. Then the momentum space period integral is*

$$\pi^{-2(\ell-1)} \int_{\mathbb{R}^{4(\ell-1)}} \frac{d^4 p_1 \cdots d^4 p_{\ell-1}}{\prod_{e \in E(G)} p_e^2}$$

2. *(position space) Choose a unit vector in $\mathbb{R}^4$ and call it 1. Choose two vertices v_0 and v_1 of G. Associate to each vertex v of G a variable x_v and set $x_{v_0} = 0$, the zero vector in $\mathbb{R}^4$ and $x_{v_1} = 1$. Orient the edges of G and given an edge e let e^+ and e^- be the initial and terminal vertices of e. Then the position space period integral is*

$$\pi^{-2(\ell-1)} \int_{\mathbb{R}^{4(\ell-1)}} \frac{\prod_{v \in V(G), v \neq v_0, v_1} d^4 x_v}{\prod_{e \in E(G)} (x_{e^+} - x_{e^-})^2}$$

3. *(parametric space) Choose an edge e_1 of G and set $a_{e_1} = 1$. Then the parametric space period integral is*

$$\int_{\substack{a_e \geq 0 \\ e \neq e_1}} \frac{\prod_{e \in E(G), e \neq e_1} da_e}{\Psi_G^2}$$

where Ψ_G is the Kirchhoff polynomial (see Definition 18).

In all cases the square of a vector denotes its norm squared.

Proof This proposition is part of Definition and Theorem 2.7 of [9] which the reader can see for details. The idea of the proof is as follows. To get from momentum space to parametric space, notice that

$$\int_0^\infty e^{-ap^2} da = \frac{1}{p^2}.$$

Use that fact on each edge in the momentum space integral to convert the integrand to

$$e^{\sum_{e \in E(G)} -a_e p_e^2}.$$

The expression in the exponent is a quadratic form and so if we do the momentum integrations first this is a big Gaussian integral. Gaussian integrals are easy to solve with standard techniques; the outcome is some appropriate power of π multiplied by a power of the determinant of the matrix of the quadratic form. For a 1-dimensional Gaussian integral the power of the determinant is $-1/2$, but because the p_e are in $\mathbb{R}^4$ we instead get $(-1/2)4 = -2$. The matrix of the quadratic form is the matrix whose (ij)th entry contains $-a_e$ if p_i and p_j appear in the momentum of e where p_i and p_j are associated to elements C_i and C_j of the cycle basis. This means that the matrix is like the graph Laplacian matrix (see Sect. 5.5) but with cycles in place of vertices. That is, it is the dual Laplacian which by the matrix tree theorem has determinant Ψ_G. This gives the $\frac{1}{\Psi_G^2}$ of the parametric integrand.

To get from position space to momentum space we need two standard facts about the Fourier transform. First, for vectors $p, x \in \mathbb{R}^4$ we have a Fourier duality

$$\int_{\mathbb{R}^4} \frac{d^4x}{(2\pi)^2} \frac{e^{ip \cdot x}}{x^2} = \frac{1}{p^2},$$

and second, the Fourier transform of 1 is the Dirac delta

$$\int_{-\infty}^\infty \frac{1}{\sqrt{2\pi}} 1 e^{-iab} db = \sqrt{2\pi} \delta(a).$$

So if we apply a Fourier transform to the position space integral we end up with factors of $\frac{1}{p_e^2}$ for each edge, with the p_e independent, and we get for each vertex v a factor

$$\int_{\mathbb{R}^4} e^{-ix_v \sum_{e \sim v} \pm p_e} d^4 v$$

where the sum runs over the edges incident to v. Treating this last integral one variable at a time we get a factor of $\delta(\sum_{e \sim v} \pm p_e)$ for each vertex—that is momentum conservation has been enforced at each vertex giving the momentum space integral.

There are a number of other equivalent integrals. The integrals given above are affine versions; projective versions exist for each of them, see [9] and [4, Sect. 3.1].

Furthermore, as well as position space, momentum space, and parametric space, there is also dual parametric space which instead of Ψ_G would use the polynomial

$$\sum_{\substack{T \text{ spanning} \\ \text{tree of } G}} \prod_{e \in T} a_e.$$

In most contexts this polynomial would be called the Kirchhoff polynomial. Here, because of the transition from momentum space, Ψ_G is more natural and hence it makes sense simply to call Ψ_G the Kirchhoff polynomial. At a graph theoretic level, these polynomials exchange spanning trees for their complements. This is the relationship between a planar graph and its dual. For non-planar graphs this is the relationship between the matroid of the graph and its matroidal dual. See Sect. 12.1 for more on duality.

Definition 20 Let G be a primitive 4-point ϕ^4 graph. The *period* of G, which we write P_G, is any of the equivalent integrals from Proposition 5. The period of an arbitrary graph is defined by the same integrals whenever they converge.

It is worth understanding how the full Feynman integral relates to the period. Suppose we begin with a scalar Feynman integral in momentum space, not yet renormalized and including masses and external momenta. If we set the masses and external momenta to 0 we almost get the period. The difference is that we have not set one of the variables to 1. Doing this takes care of the overall divergence giving a convergent integral for primitive graphs—the period.

We can also see the full Feynman integral in parametric space and relate it directly to the period in parametric space. Starting again with the momentum space Feynman integral we can follow the same procedure as for the period to convert it into parametric form. The quadratic form will now have parts which involve the masses and external momenta and so are less than quadratic in the internal momentum variables, but the part which is quadratic in the internal momenta will remain the same. The result is that there is a second Symanzik polynomial which exists in order to take care of the masses and external momenta, see for example [13]. Let's also think about the overall divergence. Suppose for the moment that we have no specified relationship between the number of edges and the number of vertices of the graph. We have not used the homogeneity of the Kirchhoff polynomial yet. Change variables so that we have a scaling variable along with all but one of the edge variables where the last edge variable has been set to 1, or any other linear constraint has been imposed (setting the sum of them to 1 is typical in the physics literature). Then we can integrate the scaling variable explicitly and what it gives is a gamma function $\Gamma(2|V(G)| - |E(G)| - 2)$. See [14, Sect. 6-2-3] for this calculation in detail. For a primitive logarithmically divergent graph this gamma function is being evaluated at a pole but the rest of the integral is convergent. Taking this rest of the integral with masses and external momenta set to 0 we get exactly the period. Thus the period is a kind of residue, or coefficient of infinity. Because of this it is largely renormalization scheme independent and is extracting a key part of the Feynman integral.

On the more mathematical side, the period of a graph is a period in the sense of Kontsevich and Zagier [15]. Namely it is an integral of an algebraic function over an algebraically defined region. Consequently it is a sensible algebro-geometric object. There has been interest [2, 4, 16–18] in taking an algebro-geometric approach to these objects. Adding to this interest is the fact that the kinds of numbers which show up in Feynman periods are very interesting. Many of them are multiple zeta values. This tells us something about the types of algebro-geometric objects which underlie them. They get much more complicated; if we move outside ϕ^4 they get arbitrarily complicated in a precise sense [19], and even within ϕ^4 they get outside of multiple zeta values [20].

The Kirchhoff polynomial is also a sensible combinatorial object, hence we can hope to get some understanding of the period with combinatorial techniques. In the denominator reduction algorithm (see Sect. 14.1) we'll see that we can, in nice cases, interpret the denominators combinatorially after integrating some of the parametric variables.

References

1. Blümlein, J., Marquard, P., Riemann, T. (eds.): Loops and Legs in Quantum Field Theory, vol. PoS(LL2014). PoS (2014)
2. Bloch, S., Kreimer, D.: Mixed Hodge structures and renormalization in physics. Commun. Number Theor. Phys. **2**, 637–718 (2008). arXiv:0804.4399
3. Broadhurst, D., Kreimer, D.: Knots and numbers in ϕ^4 theory to 7 loops and beyond. Int. J. Mod. Phys. **C6**(519–524) (1995). arXiv:hep-ph/9504352
4. Brown, F.: On the periods of some Feynman integrals. arXiv:0910.0114
5. Brown, F., Schnetz, O.: Modular forms in quantum field theory. p. 33. arXiv:1304.5342
6. Brown, F., Schnetz, O., Yeats, K.: Properties of c_2 invariants of Feynman graphs. Adv. Theor. Math. Phys. **18**(2), 323–362 (2014). arXiv:1203.0188
7. Brown, F., Yeats, K.: Spanning forest polynomials and the transcendental weight of Feynman graphs. Commun. Math. Phys. **301**(2), 357–382 (2011). arXiv:0910.5429
8. Doryn, D.: On one example and one counterexample in counting rational points on graph hypersurfaces. Lett. Math. Phys. **97**(3), 303–315 (2011). arXiv:1006.3533
9. Schnetz, O.: Quantum periods: a census of ϕ^4-transcendentals. Commun. Number Theory Phys. **4**(1), 1–48 (2010). arXiv:0801.2856
10. Panzer, E.: Feynman integrals via hyperlogarithms. In the proceedings listed as [1]. arXiv:1407.0074
11. Panzer, E.: On hyperlogarithms and Feynman integrals with divergences and many scales. J. High Energy Phys. **2014**, 71 (2014). arXiv:1401.4361
12. Crump, I., DeVos, M., Yeats, K.: Period preserving properties of an invariant from the permanent of signed incidence matrices. arXiv:1505.06987
13. Bogner, C., Weinzierl, S.: Feynman graph polynomials. Int. J. Mod. Phys. **25**(13), 2585–2618 (2010). arXiv:1002.3458
14. Itzykson, C., Zuber, J.B.: Quantum Field Theory. McGraw-Hill (1980). Dover edition (2005)
15. Kontsevich, M., Zagier, D.: Periods. In: Mathematics Unlimited–2001 and Beyond, pp. 771–808. Springer, Berlin (2001)
16. Bloch, S., Esnault, H., Kreimer, D.: On motives associated to graph polynomials. Commun. Math. Phys. **267**, 181–225 (2006). arXiv:math/0510011v1 [math.AG]

17. Bloch, S., Kreimer, D.: Feynman amplitudes and Landau singularities for 1-loop graphs. Commun. Number Theor. Phys. **4**, 709–753 (2010). arXiv:1007.0338
18. Marcolli, M.: Feynman Motives. World Scientific, Singapore (2010)
19. Belkale, P., Brosnan, P.: Matroids, motives, and a conjecture of Kontsevich. Duke Math. J. **116**(1), 147–188 (2003). arXiv:math/0012198
20. Panzer, E.: Feynman integrals and hyperlogarithms. Ph.D. thesis, Humboldt-Universität zu Berlin (2015). arXiv:1506.07243

Chapter 12
Period Preserving Graph Symmetries

Feynman graph period calculations are quite difficult, as would be expected for something capturing a substantial chunk of the interesting mathematics of the Feynman integrals themselves. So, one thing any physicist would do to try to simplify the calculations would be to look for useful symmetries. Looking for symmetries is also an excellent mathematical instinct, and particularly well suits a combinatorial perspective because the major symmetries are graph theoretic. The topic of this chapter, then, is simply, what period preserving graph transformations do we know. There may be other symmetries which we do not know.

Many of these symmetries hold more generally than stated, but to keep things straightforward and brief we will stick strictly to the 4-point ϕ^4 case and largely to the primitive case.

12.1 Planar Duality: Fourier Transform

The first period preserving graph operation we'll discuss is planar duality defined in Sect. 5.5. Broadhurst and Kreimer used this symmetry in [1] to compile their table of ϕ^4 periods. This table was key to starting number theoretic interest in Feynman periods.

We saw in Chap. 11 how the Fourier transform takes us from momentum space to position space and back. By the coincidences of 4 dimensions, there is symmetry in the Fourier transform, and so a momentum space integral can sometimes be interpreted as a position space integral for a different graph where edge cuts have been swapped for cycles. This kind of swapping is precisely planar duality of graphs and gives a graph precisely when the original graph was planar. Consequently, we get the following theorem (see Theorem 2.13 of [2] for details).

Theorem 4 *Let G be a primitive 4-point ϕ^4 graph. Suppose G is planar and G^*, its planar dual, is also a primitive 4-point ϕ^4 graph. Then $P_G = P_{G^*}$.*

© The Author(s) 2017
K. Yeats, *A Combinatorial Perspective on Quantum Field Theory*,
SpringerBriefs in Mathematical Physics 15, DOI 10.1007/978-3-319-47551-6_12

12.2 Completion

In position space, we are free to choose coordinates so as to place one vertex at the origin and another at a fixed unit vector 1. In fact there is more freedom than this. Recall, the external edges don't contribute to the period. If we imagine completing $\mathbb{R}^4$ with a point at infinity, then we can also view the external edges as each ending at the vertex at infinity with the Feynman rules just giving 1 for factors including this vertex. Then if we invert $x_i \mapsto x_i/x_i^2$ for $i \neq 0, 1, \infty$ we swap the roles of 0 and ∞ and so we can change which vertex is at infinity without changing the integral. See the end of the proof of Theorem 2.7 of [2] for details.

Inspired by the language of *completing* with a point at infinity, define the following graph operation.

Definition 21 Let G be a primitive 4-point ϕ^4 graph. Let K be the 4-regular graph given by G with a new vertex which connects to each external edge. We say K is the *completion* of G and the G is a *decompletion* of K.

See Fig. 12.1 for an example. Note that decompletions can be nonisomorphic.

By the calculations sketched above, the period is completion invariant, that is, we have the following result.

Theorem 5 *Let G and G' be primitive 4-point ϕ^4 graphs with the same completion. Then $P_G = P_{G'}$.*

This was also an important result found by Broadhurst and Kreimer in [1]. Schnetz' catalogue [2] lists graphs by their completions.

Completion is particularly interesting because it is very clear in position space, but not at all apparent parametrically. The connections of this theory to algebraic geometry and number theory are primarily through the Kirchhoff polynomial, and hence completion is quite difficult to understand algebro-geometrically or number theoretically. In particular it is still unknown if the c_2 invariant, see Chap. 15, is completion invariant.

12.3 Schnetz Twist

A third period preserving graph operation was discovered by Schnetz (see [2]). It is also best seen in position space, and is most symmetrically stated in terms of

Fig. 12.1 Two nonisomorphic decompletions of the graph K

Fig. 12.2 The Schnetz twist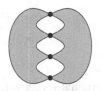

completions. Recall from Chap. 11 that a 4 regular graph has primitive decompletions if and only if it is internally 6-edge connected.

Suppose we have a 4-regular graph K with a 4-vertex cut. Let v_1, v_2, v_3, and v_4 be the cut vertices and let K_1 and K_2 be the two sides of the cut. Include in each of K_1 and K_2 the vertices v_1, v_2, v_3, and v_4 along with their incident edges which are on the correct side. We can now rejoin K_1 and K_2 identifying v_1 in K_1 with v_2 in K_2, identifying v_2 in K_1 with v_1 in K_2, identifying v_3 in K_1 with v_4 in K_2, and identifying v_4 in K_1 with v_3 in K_2. See Fig. 12.2 for an illustration. Suppose the result of this rejoining is 4-regular. Then we say the rejoined graph is related to K by a **Schnetz twist**.

Theorem 6 *Let K be an internally 6-edge connected 4 regular graph and let K' be a Schnetz twist of K which is also 4 regular. Let G and G' be decompletions of K and K' respectively. Then $P_G = P_{G'}$.*

This is Theorem 2.11 of [2]. The idea is to work in position space and set v_3 to 0 and v_4 to infinity. Then an appropriate inversion of the variables in K_2 is an invariant of the integral. In fact this works even if we move outside the 4-regular world slightly as we twist, sec [2] for details.

12.4 Products and Subdivergences

The period also has a product property. This can be stated nicely either at the completed or uncompleted level. We will start at the completed level. For the Schnetz twist we were working with a 4-vertex cut. Now suppose K is an internally 6-edge connected 4 regular graph with a 3-vertex cut.

Let v_1, v_2, and v_3 be the cut vertices and let K_1 and K_2 be the two sides of the cut, again including in each of K_1 and K_2 the cut vertices themselves along with those incident edges which are on the correct side. Now augment K_1 and K_2 each with a triangle on the vertices v_1, v_2, and v_3; call the augmented graphs $\overline{K}_1$ and $\overline{K}_2$. For an illustration see Fig. 12.3. Suppose $\overline{K}_1$ and $\overline{K}_2$ are themselves 4 regular; another way

K 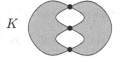 $\overline{K}_1$ $\overline{K}_2$

Fig. 12.3 Graphs involved in the completed product property

Fig. 12.4 Graphs involved in the decompleted 2-vertex cut property

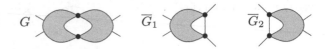

to say this is that in the original vertex cut, two edges out of each v_i went to each side. Then we have the following theorem.

Theorem 7 *With notation and hypotheses as above, let G be any decompletion of K and let G_1 and G_2 be any decompletions of $\overline{K}_1$ and $\overline{K}_2$. Then $P_G = P_{G_1} P_{G_2}$.*

See Theorem 2.10 of [2] for a proof. The idea is to calculate in momentum space with v_1, v_2, and v_3 set to 0, 1, and infinity, respectively.

Thinking at the level of primitive 4-point ϕ^4 graphs we can decomplete at v_3 to get a nice corollary. See Fig. 12.4 for an illustration.

Corollary 1 *Let G be a primitive 4-point ϕ^4 graph with a two vertex cut which partitions the external edges into two pairs. Let v_1 and v_2 be the cut vertices and let G_1 and G_2 be the two sides including the cut vertices and those incident edges which are on the correct side. Let $\overline{G}_1$ and $\overline{G}_2$ be G_1 and G_2 respectively with an additional edge from v_1 to v_2. Then $P_G = P_{\overline{G}_1} P_{\overline{G}_2}$.*

Decompleting a graph with a 3-vertex cut at a vertex other than one of the cut vertices will give a graph at the decompleted level whose period is a product but which does not have a 2-vertex cut.

Moving briefly outside the world of primitive graphs, we have not defined the period on subdivergent graphs. To do so we would need to deal with renormalization in a nontrivial way. One way to keep it polynomial based and geometric in feel is given by Brown and Kreimer in [3].

None the less, when looking for graph invariants with the same symmetries as the period we should also consider what will happen with subdivergences, or more graph theoretically, what the 4-edge cut behaviour of the invariant is. One reasonable thing which could happen is we could obtain a product, one factor for the divergent subgraph and one for the graph obtained by contracting the divergent subgraph. This corresponds to a term in the coproduct and is in some sense the leading behaviour of the Feynman integrals [3]. Another reasonable thing which could happen is to simply obtain 0 whenever there is a subdivergence.

References

1. Broadhurst, D., Kreimer, D.: Knots and numbers in ϕ^4 theory to 7 loops and beyond. Int.J.Mod.Phys. **C6**(519-524) (1995). arXiv:hep-ph/9504352
2. Schnetz, O.: Quantum periods: A census of ϕ^4-transcendentals. Communications in Number Theory and Physics **4**(1), 1–48 (2010). arXiv:0801.2856
3. Brown, F., Kreimer, D.: Angles, scales and parametric renormalization. Lett. Math. Phys. **103**, 933–1007 (2013). arXiv:1112.1180

Chapter 13
An Invariant with These Symmetries

Now we are searching for graph invariants with the symmetries of the previous chapter. One, which is inspired by flow theory, is based on permanents [1, 2]. This is the only nontrivial graph invariant I am aware of for which all the symmetries are proven. On the other hand, it is not so clear what this invariant is telling us about the Feynman period. In comparison, the c_2 invariant (see Chap. 15) is more clearly related to the period, but many of the symmetries remain conjectural.

The permanent is like the determinant except without the alternating signs.

Definition 22 Let $A = (a_{i,j})$ be an $n \times n$ matrix. The permanent of A is

$$\text{Perm}(A) = \sum_{\sigma} \prod_{i=1}^{n} a_{i,\sigma(i)}$$

where σ runs over all permutations of $\{1, 2, \ldots, n\}$.

Equivalently we can define the permanent by cofactor expansion using the same formula as for the determinant but with every term taken with a positive sign.

Unfortunately, this means the permanent is a very bad object from a linear algebra perspective. It simply doesn't behave well under linear algebra operations. Concretely, consider elementary row operations. The permanent is invariant under swapping rows and scales upon scaling rows. The problem is the row operation of adding a multiple of one row to another; call this the third row operation. In general the permanent has no nice behaviour under this operation. The first thing we need to do to define this permanent invariant is to put ourselves into a situation where we can salvage the third row operation.

Proposition 6 *Let A be a square matrix of the form*

© The Author(s) 2017
K. Yeats, *A Combinatorial Perspective on Quantum Field Theory*,
SpringerBriefs in Mathematical Physics 15, DOI 10.1007/978-3-319-47551-6_13

$$A = \begin{bmatrix} B \\ B \\ \vdots \\ B \end{bmatrix}$$

where there are n copies of the matrix B. Then Perm(A) is unchanged modulo n + 1 if we perform the same third row operation in each B block.

This is Lemma 5 of [2]. The idea is to expand by linearity in the changed rows and note that the parts coming from adding the multiple of another row give matrices with $n + 1$ rows the same. Any permutation of these $n + 1$ rows leaves the matrix and hence the permanent unchanged, so the permanent for these matrices must be divisible by $(n + 1)!$ and hence is 0 modulo $n + 1$.

The point is that for these kinds of block matrices, as long as we work modulo $n+1$, we do get a well behaved linear algebra object. To apply this to a graph invariant recall the signed incidence matrix from Sect. 5.5 and let E be the signed incidence matrix with any one row removed. For a 4-point ϕ^4 graph, this matrix will be $\ell \times 2\ell$ where ℓ is the loop number. Consequently

$$\mathrm{Perm}\left(\begin{bmatrix} E \\ E \end{bmatrix}\right) \quad \mathrm{mod}\ 3$$

is an object of the desired sort. There is one subtlety, namely the choices involved in building E can change the overall sign and so we define the **graph permanent** in work with Crump and DeVos [2], to be this permanent up to sign. In fact for any prime $p > 2$, we can stack $p - 1$ copies of E vertically and $(p - 1)/2$ horizontally to obtain a square matrix of the required type. Taking the permanent modulo p and up to sign in that case gives the **extended graph permanent at** p defined in [1] by Crump. The overall **extended graph permanent** is the sequence indexed by odd primes of the extended graph permanent at each prime.

The extended graph permanent has the same graph symmetries as the period [1]. To keep things simpler, we'll just look at the arguments in the original graph permanent ($p = 3$) case; the extended graph permanent arguments are similar.

For planar duality, since we can row reduce we can put E into the form $[I\ A]$. The matrix $[-A^T\ I]$ is row equivalent to the E matrix for the dual (see [3] for this fact in the more general matroid language). By cofactor expanding the I parts first for both matrices, and since the permanent is transpose invariant, the graph permanents agree. For details see [2, Proposition 20].

For completion and decompletion we need to think about how to index the nonzero terms in the cofactor expansion of the permanent. A term in the cofactor expansion is a choice of entries with exactly one in each row and in each column. Each nonzero entry in the matrix is a pair of a vertex and an edge; think of it as a half edge. One entry in each column means that each edge is involved once. One entry in each row means that each vertex other than the one corresponding to the removed row is taken twice (once for each block). We will call the vertex corresponding to the removed row the **special vertex**. So a nonzero entry in the matrix is a choice of half edges of the graph with one half edge from each edge and two half edges around each vertex except for none around the special vertex. We can extend this selection to the completed graph by taking all four half edges around the completion vertex. If for all edges in the completed graph we swap which half edge we choose then we have swapped the role of special vertex and decompletion vertex and obtained a term in the cofactor expansion for this swapped graph. This gives a bijection between the terms in the expansions of the two permanents and hence they are the same up to sign. For details see [2, Theorem 16].

For the Schnetz twist we use the same idea as for completion. On the completed graphs, set one of the cut vertices to be to be the decompletion vertex and one of the others to be the special vertex. Again think of the terms in the cofactor expansion of the permanent as selections of half edges, but this time swap only one side of the cut to get a term for the twisted graph. For details see [2, Proposition 19].

For the product, by completion invariance we can choose to decomplete at one of the vertices of the 3-vertex cut and so it suffices to consider decompleted graphs with a 2-vertex cut. Of those two vertices, make one of them the special vertex and cofactor expand along the other to get the desired identity. For details see [2, Theorem 22].

Furthermore, subdivergences also give products. This comes from cofactor expanding along the four cut edges, see [2, Theorem 24].

These symmetries suggest the extended graph permanent should be a period invariant in the sense that if two graphs have the same Feynman period then they should have the same extended graph permanent. Hence the extended graph permanent should be telling us something about the period. In the case of the original graph permanent, because of the sign ambiguity, we have a binary invariant, zero or nonzero. Unfortunately it is not at all clear what it is telling us about the period. The extended graph permanent is even more tantalizing. It contains more information and has some mysterious connections with the c_2 invariant but it is still unclear what it tells us about the Feynman period (see [1]).

References

1. Crump, I.: Properties of the extended graph permanent. arXiv:1608.01414
2. Crump, I., DeVos, M., Yeats, K.: Period preserving properties of an invariant from the permanent of signed incidence matrices. arXiv:1505.06987
3. Oxley, J.: Matroid Theory. Oxford University Press, Oxford (1992)

Chapter 14
Weight

14.1 Denominator Reduction

In [1] Brown gave a technique for calculating Feynman periods and Feynman integrals of certain graphs by step by step integration in parametric space. We will return to this algorithm in Chap. 16. For now, we will investigate a key part of the algorithm known as denominator reduction. Denominator reduction is about keeping track of the denominators which show up over the course of the integration algorithm. These denominators are, like the Kirchhoff polynomial, polynomials which can be understood combinatorially. Furthermore they contain important information about the period as a whole. In particular they know about the weight which we will define in Sect. 14.2.

First we need the building blocks for the polynomials.

Definition 23 Let G be a graph. Let

$$M = \begin{bmatrix} \Lambda & E^T \\ -E & 0 \end{bmatrix}$$

be as in Sect. 5.5. Let I and J and K be sets of edge indices of G with $|I| = |J|$. Let $M(I, J)$ be the matrix obtained from M by removing the rows indexed by I and the columns indexed by J. Then the polynomial

$$\Psi_{G,K}^{I,J} = \det M(I, J)|_{\substack{a_i=0 \\ i \in K}}$$

is called a *Dodgson polynomial*.

Dodgson polynomials satisfy a contraction-deletion relation.

Proposition 7 *For $i \notin I \cup J \cup K$,*

© The Author(s) 2017
K. Yeats, *A Combinatorial Perspective on Quantum Field Theory*,
SpringerBriefs in Mathematical Physics 15, DOI 10.1007/978-3-319-47551-6_14

Fig. 14.1 An example for a
spanning forest polynomial

$$\Psi_{G,K}^{I,J} = \Psi_{G,K}^{I\cup\{i\},K\cup\{i\}} a_i + \Psi_{G,K\cup\{i\}}^{I,J}$$
$$\Psi_{G,K}^{I\cup\{i\},K\cup\{i\}} = \Psi_{G\setminus i,K}^{I,K}$$
$$\Psi_{G,K}^{I\cup\{i\},K\cup\{i\}} = \Psi_{G/i,K}^{I,K}$$

Proof These follow from the form of the matrix defining the Dodgson polynomials.
(See for example [2, Section 2.2]).

Dodgson polynomials can also be understood, via the all-minors matrix-tree theorem [3], as sums of spanning forest polynomials.

Definition 24 Let P be a set partition of a subset of the vertices of a graph G. Define

$$\Phi_G^P = \sum_F \prod_{e\notin F} a_e$$

where the sum runs over spanning forests F of G with a bijection between the trees
of F and the parts of P where each vertex in a part lies in its corresponding tree.

Note that trees consisting of isolated vertices are permitted. For example in the graph
given in Fig. 14.1 if the partition has the two white vertices in one part and the black
vertex in the other part then the spanning forest polynomial is

$$ab(ec + cd + cf + df).$$

The relationship between Dodgson polynomials and spanning forest polynomials
is given in [4].

Proposition 8 (Proposition 12 from [4]) *Let I, J, K be sets of edge indices of G
with $|I| = |J|$, then*
$$\Psi_{G,K}^{I,J} = \sum_P \pm \Phi_{G\setminus(I\cup J\cup K)}^P$$

*where the sum runs over all set partitions P of the end points of edges of $(I \cup J \cup
K) \setminus (I \cap J)$ with the property that all the forests corresponding to P become trees
in both $G \setminus I/(J \cup K)$ and $G \setminus J/(I \cup K)$.*

Now let's return to the Dodgson polynomials in order to define denominator
reduction.

Proposition 9 *Given 5 edge indices i, j, k, l, m*

$$^5\Psi_G(i, j, k, l, m) = \pm \left(\Psi_{G,m}^{ij,kl} \Psi_G^{ikm,jlm} - \Psi_{G,m}^{ik,jl} \Psi_G^{ijm,klm} \right)$$

is independent (up to overall sign) of the order of i, j, k, l, m.

For a proof see Lemma 87 in [1]. We call $^5\Psi_G(i, j, k, l, m)$ the **5-invariant** of G depending on edges i, j, k, l, m. The denominator reduction story starts with the 5-invariant because fewer integrations do not have the general shape yet and more integrations are not always possible.

Definition 25 Given G with at least 5 edges, **denominator reduction** is a sequence of polynomials $D_5, D_6, \ldots D_k$, defined by

- $D_5(i_1, i_2, i_3, i_4, i_5) = {}^5\Psi_G(i_1, i_2, i_3, i_4, i_5)$
- If $D_j(i_1, \ldots, i_j)$ can be factored as

$$D_j(i_1, \ldots, i_j) = (Aa_t + B)(Ca_t + D)$$

  where A, B, C, D are polynomials (not necessarily nonzero but not all zero) in the edge variables not involving a_t then $D_{j+1}(i_1, \ldots, i_j, t) = \pm(AD - BC)$
- If $D_{j+1} = 0$ or D_j cannot be factored then denominator reduction ends.

Denominator reduction is important because it is a purely algebraic process but it gives the denominators remaining in the parametric Feynman period integral after successive parametric integrations.

There are a few things to notice about denominator reduction. First, each $D_j(i_1, \ldots, i_j)$ is defined up to overall sign. Second, for a given graph, different edge orders will give a different sequence of polynomials, and such sequences may not all be the same length. Long sequences are better for our purposes because the longer the sequence the further the integration algorithm works and the more we can learn combinatorially about the period. Third, the factorizations in the denominator reduction algorithm are often combinatorially meaningful in the sense that the factors are themselves sums of Dodgson polynomials, hence of spanning forest polynomials.

Fourth, for many purposes we can begin the sequence with D_4 or even D_3 at the expense of no longer having a unique choice. For any choice of four edges i, j, k, l there are three generically distinct possible choices for D_4,

$$\Psi^{ij,kl}\Psi^{ik,jl}, \quad \Psi^{ij,kl}\Psi^{il,jk}, \quad \text{or} \quad \Psi^{ik,jl}\Psi^{il,jk}$$

any of which we will call D_4 and each of which yields the same D_5 and onwards following the denominator reduction algorithm. For D_3 we can take any permutation of indices of

$$\Psi_k^{i,j}\Psi^{ik,jk}.$$

14.2   Weight Drop and Double Triangles

Feynman periods give interesting and presumably transcendental numbers so it is
natural to ask about their transcendental weight. Intuitively, the transcendental weight
of a number is the minimum number of nested integrals needed to write it as a period,
that is, as an integral of an algebraic function over an algebraically defined region.
Unfortunately, this intuitive definition is not well suited to upgrading to a formal
definition, so we will have to be satisfied with some special cases.

To begin with, we find many multiple zeta values among the Feynman periods.
For low loop orders they suffice (see the tables of [5]). If $s_1, s_2, \ldots, s_k$ are positive
integers with $s_1 > 1$ then the **multiple zeta value** $\zeta(s_1, s_2, \ldots, s_k)$ is defined to be

$$\zeta(s_1, s_2, \ldots, s_k) = \sum_{n_1 > n_2 > \cdots > n_k > 0} \frac{1}{n_1^{s_1} n_2^{s_2} \cdots n_k^{s_k}}.$$

These generalize special values of the Riemann zeta function and are mathematically
very interesting, see [6–8]. For our purposes we don't need to know any more about
them beyond the definition of the **weight** of a multiple zeta value as the sum $s_1 +
\cdots + s_k$. This is consistent with the intuitive idea of weight because multiple zeta
values also have an iterated integral representation of the appropriate length. All
nontrivial multiple zeta values are believed to be transcendental and there are precise
conjectures about the algebraic relations between them. However, it is extremely
difficult to prove transcendence, or even to prove irrationality, of these numbers.

For more general periods we are left without a good definition of weight to work
with. From a mathematical perspective a good approach is to upgrade periods to
mixed Hodge structures and use the weight filtration there, but then, outside of
simple examples, we are left without good, concrete tools.

Instead we will use the special properties of Feynman periods to get partial combinatorial access to the weight using denominator reduction. Integrating one variable at
a time in parametric space following Brown [1], we schematically get the following:

- Begin with $\int \frac{1}{\Psi_G^2}$.
- Integrate one edge variable to get $\int \frac{1}{\Psi_{G/e_1} \Psi_{G \setminus e_1}}$.
- Integrate another edge variable to get $\int \frac{\text{logs}}{(\Psi^{1,2})^2}$.
- Integrate a third edge variable to get $\int \frac{\text{more logs}}{\text{a polynomial}}$.
- Integrate a fourth edge variable to get $\sum \int \frac{\text{dilogs}}{D_4}$, one term for each version of D_4.
  Along with the actual dilogs are products of pairs of logs which are also weight 2.
- Integrate a fifth edge to get $\int \frac{\text{weight 3 stuff}}{5\psi}$.

In all cases the pieces are fully explicit, see [1] or [4, Sect. 4.1] for further details.
In particular the pieces are made up of polylogarithms which we will look at briefly
in Chap. 16. Polylogarithms have multiple zeta values as special evaluations and
have a comparable definition of weight. From the point of view of weight, what is
happening in the sketch above is that the first and third integrations do not change the

weight while the other integrations each increase the weight by 1. The cases where the weight was unchanged were the cases where the denominator ended up being a square after the previous step.

Schematically, Brown's algorithm looks similar in general. From $\int \frac{\text{weight } n \text{ stuff}}{D_n}$,

- If D_n factors into two distinct factors then we get $\int \frac{\text{weight } n + 1 \text{ stuff}}{D_{n+1}}$.
- If D_n is a square we get a sum of pieces of weight at most n in the numerator.
- If D_n doesn't factor then the algorithm fails.

Again the pieces are all explicit. We will see a bit more about them in Chap. 16 and [1] has further details.

Recall that for a 4-point ϕ^4 graph the number of edges is 2ℓ where ℓ is the loop number of the graph and one of the edge variables is not integrated in the period. We see from the above that two of the integrations don't increase the weight, so this leaves $2\ell - 3$ possible integrations, so the maximal weight of a Feynman period obtained in this way is $2\ell - 3$. From the known multiple zeta value Feynman periods we know this maximal weight is frequently achieved. On the other hand, if we ever hit the square denominator case for some D_n then the weight will be less than $2\ell - 3$. In this case we say the graph has **weight drop**. We can use this as a purely combinatorial definition of weight drop.

The easiest weight drop graphs to understand are graphs which are products in the sense of Sect. 12.4. Suppose G is a primitive 4-point ϕ^4 graph with a 2 vertex cut separating the external edges so $P_G = P_{G_1} P_{G_2}$. Say G has loop number ℓ and G_1 has loop number k. In splitting G into G_1 and G_2, we break one loop by splitting at the cut vertices and adding the extra edge back into G_1 and G_2 adds one loop to each of them. So the loop number of G_2 is $\ell - k + 1$. Then the maximum weight of G_1 is $2k - 3$ and the maximum weight of G_2 is $2(\ell - k + 1) - 3$. The weight of a product is the sum of the weights of the pieces so the maximum weight of G is $2k - 3 + 2(\ell - k + 1) - 3 = 2\ell - 4 < 2\ell - 3$ so G has weight drop.

A really interesting example of the graph theory telling us about the periods gives another criteria for weight drop. Suppose G is of the form

$$G = $$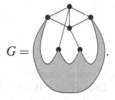

Let H be the graph which is otherwise unchanged from G but the pair of triangles is converted to a single triangle in the following way

$$H = $$ 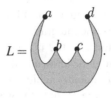 .

In G we can denominator reduce the 7 edges shown explicitly and obtain

$$D_7 = \Phi_L^{\{a,d\},\{b\},\{c\}} \left(\Phi_L^{\{a,b\},\{c,d\}} - \Phi_L^{\{a,c\},\{b,d\}} \right)$$

where

$$L = $$

In H we can denominator reduce the 5 edges shown explicitly and obtain

$$D_5 = \Phi_L^{\{a,d\},\{b\},\{c\}} \left(\Phi_L^{\{a,b\},\{c,d\}} - \Phi_L^{\{a,c\},\{b,d\}} \right)$$

which is the same. The calculations can all be done at the level of spanning forest polynomials where they are purely graph theoretic manipulations (see [4, Theorem 35]). The point is that the denominators are the same after that point, so the denominators will continue to agree for any further denominator reductions. In particular the denominator reduction will give a square in one case if and only if it gives a square in the other, so G has weight drop if and only if H has weight drop.

In the next chapter we will look at an arithmetic graph invariant which can see the weight of the period.

References

1. Brown, F.: On the periods of some Feynman integrals. arXiv:0910.0114
2. Brown, F., Schnetz, O.: A K3 in ϕ^4. Duke Math J. **161**(10), 1817–1862 (2012). arXiv:1006.4064
3. Chaiken, S.: A combinatorial proof of the all minors matrix tree theorem. SIAM J. Algebraic Discrete Methods **3**(3), 319–329 (1982)
4. Brown, F., Yeats, K.: Spanning forest polynomials and the transcendental weight of Feynman graphs. Commun. Math. Phys. **301**(2), 357–382 (2011). arXiv:0910.5429
5. Schnetz, O.: Quantum periods: a census of ϕ^4-transcendentals. Commun. Number Theory Phys. **4**(1), 1–48 (2010). arXiv:0801.2856
6. Goncharov, A.: Multiple ζ-values, Galois groups, and geometry of modular varieties. In: Arbeitstagung (1999). arXiv:math.AG/0005069

7. Hoffman, M.E.: Algebraic aspects of multiple zeta values. In: Aoki, T., Kanemitsu, S., Nakahara, M., Ohno, Y. (eds.) Zeta Functions, Topology and Quantum Physics, vol. 14 in Developments in Mathematics, pp. 51–74. Springer (2005). arXiv:math.QA/0309425
8. Zudilin, W.: Algebraic relations for multiple zeta values. Russ. Math. Surv. **58**(1), 1–29 (2003)

... ...
... ...

Chapter 15
The c_2 Invariant

Schnetz [1] defined the following graph invariant based on counting points on the affine variety defined by the vanishing of Ψ_G.

Definition 26 Let p be a prime, let $\mathbb{F}_p$ be the field with p elements, and let G have at least 3 vertices. Let $[\Psi_G]_p$ be the number of points on the affine variety of Ψ_G over $\mathbb{F}_p$. Define the c_2-invariant of G at p to be

$$c_2^{(p)}(G) = \frac{[\Psi_G]_p}{p^2} \mod p$$

Write $c_2(G)$ for the sequence $(c_2^{(2)}(G), c_2^{(3)}(G), \ldots)$.

The fact that this is well defined depends on G having at least three vertices (see Proposition 2 of [2]). When $c_2^{(p)}(G)$ is a constant k independent of p we'll just write $c_2(G) = k$.

To make this definition more motivated we need to understand a bit of the history. From the calculations of [3] there came the conjecture that all ϕ^4 Feynman periods could be written in terms of linear combinations of products of multiple zeta values. Thinking algebro-geometrically, one is led to think that if this conjecture were true, then these multiple zeta values should be appearing for some good reason. In the language of motives, this good reason should be that the varieties defined by Kirchhoff polynomials give mixed Tate motives. Coming back to the concrete side, this would mean that the point counting functions for these varieties would be polynomials in p. Likely inspired by thoughts along these lines, Kontsevich informally conjectured this last part in 1997. This conjecture turned out to be false. This was proved in the context of all graphs by Belkale and Brosnan [4], and explicit ϕ^4 counterexamples were found a little later [2, 5].

However, there is more to get out of this idea of the polynomiality of point count functions of Kirchhoff varieties. In the cases when the point count function is a

© The Author(s) 2017
K. Yeats, *A Combinatorial Perspective on Quantum Field Theory*,
SpringerBriefs in Mathematical Physics 15, DOI 10.1007/978-3-319-47551-6_15

polynomial the c_2 invariant is the quadratic coefficient of this polynomial—this is the reason for the name c_2. In particular, the c_2 invariant in such cases is constant as a function of p. In other cases the c_2 invariant is nonconstant as a function of p but it is still informative and perhaps even more interesting as one gets coefficient sequences of modular forms [6, 7] and even more mysterious sequences. The c_2 invariant is telling us about what kinds of numbers show up in the Feynman period.

We know some things about the symmetries of the c_2 invariant. Let G be a primitive 4-point graph in ϕ^4. One key fact is that the c_2 invariant can be computed using denominator reduction. Specifically, for $n \geq 3$,

$$c_2^{(p)}(G) = (-1)^n [D_n]_p \quad \mod p.$$

This is Theorem 29 of [2] and is proved by stepping through the denominator reduction process keeping track of the point counts. This has many useful consequences ([2]):

• If G has weight drop in the sense of Chap. 14 then it has $c_2(G) = 0$. In particular, if G has a 2-vertex cut then $c_2(G) = 0$.
• If H differs from G by a double triangle transformation (see Sect. 14.2) then $c_2^{(p)}(G) = c_2^{(p)}(H)$ for all p.

With a bit more work, denominator reduction is also useful for showing that ϕ^4 graphs with subdivergences all have $c_2(G) = 0$ (see [8]).

This still leaves many of the major symmetries of Chap. 12 open. All of them are fully supported by all the examples which have been calculated. Doryn has proved that the c_2 invariant is invariant under planar duality (Corollary 34 and Theorem 5 of [9]) but not necessarily under a more general duality for nonplanar graphs. The key idea is to look at the analogue of the c_2 invariant in dual parametric space as well as in momentum space and position space and try to show that they coincide. Completion invariance for c_2 is open (Conjecture 4 of [2]) as is the Schnetz twist.

Brown and Schnetz also make a stronger conjecture (Conjecture 5 of [2]), that $P_{G_1} = P_{G_2}$ implies $c_2(G_1) = c_2(G_2)$.

Brown and Schnetz [6] have computed c_2 invariants systematically on primitive 4-point ϕ^4 graphs up to 9 loops on the first 12 primes. They found c_2 invariants which were constant, c_2 invariants which were constant except for a dependence on the modulo class of the prime, c_2 invariants which were the coefficient sequences of modular forms, and some unknown sequences. They computed the unknown sequences up to 50 or 100 primes. These sequences give great staring material for people who like sequences, (see [6]).

Rather than fixing a graph and calculating the c_2 invariant for different primes, a slightly different approach is to fix a prime and calculate the c_2 invariant at this prime on different graphs in some structured family. This was looked at in [10] for decompletions of circulant graphs. The outcome is (often large) recurrences so the flavour is quite different. The techniques are largely graph theoretic working with spanning forest polynomials.

References

1. Schnetz, O.: Quantum field theory over $\mathbb{F}_q$. Electr. J. Combin. **18** (2011). arXiv:0909.0905
2. Brown, F., Schnetz, O.: A K3 in ϕ^4. Duke Math J. **161**(10), 1817–1862 (2012). arXiv:1006.4064
3. Broadhurst, D., Kreimer, D.: Knots and numbers in ϕ^4 theory to 7 loops and beyond. Int. J. Mod. Phys. **C6**(519–524) (1995). arXiv:hep-ph/9504352
4. Belkale, P., Brosnan, P.: Matroids, motives, and a conjecture of Kontsevich. Duke Math. J. **116**(1), 147–188 (2003). arXiv:math/0012198
5. Doryn, D.: On one example and one counterexample in counting rational points on graph hypersurfaces. p. 91. arXiv:1006.3533
6. Brown, F., Schnetz, O.: Modular forms in quantum field theory. p. 33. arXiv:1304.5342
7. Logan, A.: New realizations of modular forms in Calabi-Yau threefolds arising from ϕ^4 theory. arXiv:1604.04918
8. Brown, F., Schnetz, O., Yeats, K.: Properties of c_2 invariants of Feynman graphs. Adv. Theor. Math. Phys. **18**(2), 323–362 (2014). arXiv:1203.0188
9. Doryn, D.: The c_2 invariant is invariant. arXiv:1312.7271
10. Yeats, K.: A few c_2 invariants of circulant graphs. p. 33. arXiv:1507.06974

Chapter 16
Combinatorial Aspects of Some Integration Algorithms

The art of computing Feynman integrals has always involved graph theory in the sense that the specific structure of each Feynman graph really matters. Feynman integration is very hard so quantum field theorists have become very skilled at extracting every bit of information they can from the structure of the graphs as well as having many more analytic tricks. Recently there have been some new ideas which are related to this brief. This chapter will overview some of their combinatorial aspects.

We have already looked a bit at Brown's denominator reduction algorithm in Chap. 14. The analytic piece that we have ignored so far has been the polylogarithms which appear in the numerators. Let $\sigma_1, \ldots, \sigma_n$ be n distinct points in $\mathbb{C}^*$ and let $x_0, x_1, \ldots, x_n$ be $n + 1$ letters. Given a word $w = x_{i_1} \cdots x_{i_r}$ on these letters build an iterated integral

$$L_w(z) = \int_{0 \leq t_r \leq t_{r-1} \leq \cdots \leq t_1 \leq z} \frac{dt_r}{t_r - \sigma_{i_r}} \cdots \frac{dt_1}{t_1 - \sigma_{i_1}}$$

where $\sigma_0 = 0$. Then for $i \neq j$ we get the following indefinite integrals with constants of integration omitted

$$\int \frac{L_w(z)}{z - \sigma_i} dz = L_{x_i w}$$

$$\int \frac{L_w(z)}{(z - \sigma_i)(z - \sigma_j)} dz = \frac{1}{\sigma_i - \sigma_j} (L_{x_i w}(z) - L_{x_j w}(z))$$

$$\int L_{x_{i_1} \cdots x_{i_n}}(z) dz = \sum_{k=1}^{n} (-1)^k (z - \sigma_{i_k}) L_{x_{i_{k+1}} \cdots x_{i_n}}(z)$$

$$\int \frac{L_{x_i^r x_j w}(z)}{(z - \sigma_i)^2} dz = \frac{1}{\sigma_i - \sigma_j} (L_{x_i w}(z) - L_{x_j w}(z)) + \frac{1}{1 - \sigma_i} \sum_{k=1}^{r} (-1)^{k+1} L_{x_i^{r-k} x_j w}.$$

© The Author(s) 2017
K. Yeats, *A Combinatorial Perspective on Quantum Field Theory*,
SpringerBriefs in Mathematical Physics 15, DOI 10.1007/978-3-319-47551-6_16

See [1, Sect. 4.1] for details. We can see in these formulas the different denominator reduction steps. The second formula is the case when the denominator factors into distinct factors. After normalizing appropriately we end up with the new denominator built of the old ones and in the numerator the iterated integrals are of weight one larger than before. The fourth formula is the case when the denominator is a square. The result is hairier but all of the iterated integrals on the right hand side have the same weight or lower.

Polylogarithms have many analytic and geometric properties and are very rich objects. However, they are substantially controlled by the word w. Denominator reduction accentuates this in the way the words are built letter by letter. In some sense what this algorithm does is convert from graphs to words in a subtle and physically significant way. This algorithm is readily implementable; variants and extensions have been implemented by Bogner [2] and Panzer [3–5].

In [6] Schnetz introduced *graphical functions*. A graphical function is the evaluation of the Feynman integral of a graph G with three distinguished vertices 0, 1, z using position space Feynman rules and with no integration over z [6, Sect. 1.3]. The unintegrated z behaves like a hook; more edges can be added connecting at 0, 1 and z and the integration continues. Thus graphical functions give a recursive way to compute certain classes of Feynman integrals again using polylogarithms.

The most important class computed in this way is the zigzag family. Zigzag graphs are graphs of the form showed in the first part of Fig. 16.1. Broadhurst and Kreimer conjectured in [7] that the period of the zigzag graph with ℓ loops is a particular rational multiple (see below) of $\zeta(2\ell - 3)$. Brown and Schnetz prove this in [8] using graphical functions. Zigzag graphs are well suited to integration by graphical functions. Work with the planar dual, see the second part of Fig. 16.1. Take z as running along the middle with the top vertex as 0 and the bottom vertex as 1. So to calculate the period by graphical functions start with the path from 0 to z to 1. Then step along adding a new edge out of z, integrating the old z and putting the new z at the other end of this edge, and alternately adding an edge from z to 1 or an edge from z to 0. With the help of some identities of multiple zeta values, the general form of the zigzag period was proved by this approach in [8] to be

$$\frac{4(2\ell - 2)!}{\ell!(\ell - 1)!} \left(1 - \frac{1 - (-1)^\ell}{2^{2\ell-3}} \right) \zeta(2\ell - 3).$$

Fig. 16.1 A zigzag graph and its planar dual

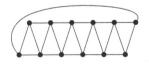 

Ultimately, many tools, including combinatorial ones, come together to make these integration algorithms work. This is the same phenomenon which has appeared throughout this brief: combinatorics playing a nontrivial role in the physics and mathematics of quantum field theory.

References

1. Brown, F., Yeats, K.: Spanning forest polynomials and the transcendental weight of Feynman graphs. Commun. Math. Phys. **301**(2), 357–382 (2011). arXiv:0910.5429
2. Bogner, C.: MPL—a program for computations with iterated integrals on moduli spaces of curves of genus zero. Comput. Phys. Commun. **203**, 339–353 (2016)
3. Panzer, E.: Feynman integrals via hyperlogarithms. In the proceedings listed as [1]. arXiv:1407.0074
4. Panzer, E.: On hyperlogarithms and Feynman integrals with divergences and many scales. J. High Energ. Phys. **2014**, 71 (2014). arXiv:1401.4361
5. Panzer, E.: Feynman integrals and hyperlogarithms. Ph.D. thesis, Humboldt-Universität zu Berlin (2015). arXiv:1506.07243
6. Schnetz, O.: Graphical functions and single-valued multiple polylogarithms. Commun. Number Theory Phys. **8**(4), 589–675 (2014). arXiv:1302.6445
7. Broadhurst, D., Kreimer, D.: Knots and numbers in ϕ^4 theory to 7 loops and beyond. Int. J. Mod. Phys. **C6**(519–524) (1995). arXiv:hep-ph/9504352
8. Brown, F., Schnetz, O.: Single-valued multiple polylogarithms and a proof of the zig-zag conjecture. J. Number Theor. **148**, 478–506 (2015). arXiv:1208.1890

Index

© The Author(s) 2017
K. Yeats, *A Combinatorial Perspective on Quantum Field Theory*,
SpringerBriefs in Mathematical Physics 15, DOI 10.1007/978-3-319-47551-6

Printed in the United States
By Bookmasters